On Quantum Things

On Quantum Things

A Constructive Guide to the Fundamentals of Quantum Theory

Valia Allori

BLOOMSBURY ACADEMIC
NEW YORK • LONDON • OXFORD • NEW DELHI • SYDNEY

BLOOMSBURY ACADEMIC
Bloomsbury Publishing Inc, 1359 Broadway, New York, NY 10018, USA
Bloomsbury Publishing Plc, 50 Bedford Square, London, WC1B 3DP, UK
Bloomsbury Publishing Ireland, 29 Earlsfort Terrace, Dublin 2, D02 AY28, Ireland

BLOOMSBURY, BLOOMSBURY ACADEMIC and the Diana logo
are trademarks of Bloomsbury Publishing Plc

First published in the United States of America 2026

Library of Congress Cataloging-in-Publication Data

Names: Allori, Valia author
Title: On quantum things / Valia Allori.
Description: New York : Bloomsbury Academic, 2026. |
Includes bibliographical references. | Summary: "A realist, particle-based view of quantum theory-where the wavefunction no longer plays a central role-offers a clear, commonsense alternative to dominant metaphysical views"– Provided by publisher.
Identifiers: LCCN 2025032830 (print) | LCCN 2025032831 (ebook) | ISBN 9781498581684 hardback | ISBN 9798216375470 paperback | ISBN 9781978764583 epub | ISBN 9798216267355 pdf Subjects: LCSH: Quantum theory
Classification: LCC QC174.12 .A455 2026 (print) | LCC QC174.12 (ebook)
LC record available at https://lccn.loc.gov/2025032830
LC ebook record available at https://lccn.loc.gov/2025032831

ISBN: HB: 978-1-4985-8168-4
ePDF: 979-8-216-26735-5
eBook: 978-1-9787-6458-3

Typeset by Integra Software Services Pvt. Ltd.

For product safety related questions contact productsafety@bloomsbury.com.

To find out more about our authors and books visit www.bloomsbury.com
and sign up for our newsletters.

To
Lara and Morena
Leo and Ale

Contents

Figures

Acknowledgments

I am particularly grateful to all my colleagues and friends who, over the years, have given me feedback on the ideas developed in this book. Their invaluable help, insightful and fair comments, and challenging objections have been essential to this work. I do not think it makes sense to name everyone, as I would undoubtedly forget someone. However, there are a few individuals without whom this book would not have existed (at least, not in its current form). I would like to thank Jean Bricmont and Alyssa Ney, to whom I am deeply indebted for their genuine friendship, unwavering support, and for never growing tired of debating these issues with me. Their countless sharp comments and insights (always from opposing perspectives!) have greatly helped me better understand different viewpoints and have guided me in shaping and refining my arguments. I would also like to express my gratitude to Mario Hubert, Peter Lewis, Barry Loewer, Wayne Myrvold, and Travis Norsen, who offered invaluable feedback, both written and verbal, on the material developed in this book and the book itself. Additionally, I am deeply appreciative of the help, support, and thoughtful feedback I have received over the years from David Albert, Craing Callender, Eddy Chen, Sheldon Goldstein, Tim Maudlin, Jill North, Lev Vaidman, David Wallace, and Nino Zanghì.

In this book, I present and discuss my own views while also summarizing and engaging with the views of others, as I best understand them. All figures and illustrations were created by myself. Any errors or misrepresentations—conceptual, factual, or visual—are entirely my own.

Prologue

This is a relatively short book about the metaphysics of quantum theory, in which I present my view on what we should believe exists in the world if quantum theory is true. I endorse scientific realism, the view that scientific theories can inform us about the nature of reality. The title indicates that the book is about "quantum things," referencing Lucretius' *De Rerum Natura* (*On the Nature of Things*), arguably the first text on naturalized metaphysics. Also, the subtitle carries a double meaning, relying on the ambiguity of the word *fundamentals*. The term can refer to the basic principles of a discipline, as in a title like *The Fundamentals of Tensor Calculus for Physicists*. However, it can also be understood as *the fundamental entities of the quantum world*.

In the first three chapters, I explore the former meaning (the basic principles), offering a brief review of several approaches that examine the connection between physics and metaphysics, and how these have been adapted to address quantum physics. In the later chapters, I turn to the second meaning of the subtitle (what lies at the foundation). There, I propose what I consider to be the best approach for investigating the metaphysics of quantum theory, and offer my answer to the question: What are the fundamental ingredients of the quantum world?

A Short Overview

Chapter 1 introduces scientific realism, or simply *realism*. Roughly put, this view holds that we are justified in believing that a successful theory is (approximately) true. I begin by discussing the most compelling argument for realism: namely, that a theory's success is best explained by its truth. If realism were false, it would be a miracle that the world behaves as if electrons exist, even though we would not be justified in believing that they do.

While realism might seem like a commonsense position, it has historically faced two serious challenges, both predating quantum theory. The first is *underdetermination*: empirical data can be compatible with an infinite number of metaphysically incompatible theories, and we have no reliable method to determine which one is true. The second is the *pessimistic meta-induction*: since many past successful theories have turned out to be false, the success of a theory cannot be taken as a reliable indicator of its truth. Realists have responded to the first challenge by arguing that there is more to a theory than empirical adequacy. Some theories are simpler, more explanatory, or more fruitful than others, and should therefore be preferred. To clarify the notion of explanatory power, I distinguish between *constructive* and *principle* theories: the former offer compositional, dynamical explanations of phenomena, while the latter constrain phenomena using general principles. What I refer to as the *classical schema of explanation* regards constructive theories as providing deeper explanations, particularly when they are grounded in symmetries, which help ensure that the explanations are as perspective-independent as possible. In response to the second challenge, realists have suggested narrowing realism to those entities responsible for the success of past theories, provided that such entities persist through theory change. One way to do this is what I call *explanationism*, which distinguishes between a theory's *presuppositional* and *working* posits. According to this view, we are justified in believing in the existence of the entities represented by the working posits, as they are responsible for the theory's success. Another promising strategy is *structural realism*, which claims that what carries over from one successful theory to another, and thus what we should be realist about, is structure. The most radical form of this view is eliminative structuralism, or *eliminativism*, which denies that we are justified in believing in fundamental objects at all. Instead, structure alone is taken to be ontologically significant. This view gained traction with the advent of quantum theory. Initially, it was believed that quantum mechanics undermines realism, because it seems to assign contradictory properties to objects, for example, a particle being both "here" and "there" simultaneously. These *superpositions* extend to all levels of description, including macroscopic objects. This issue is known as the *measurement problem*, and it arises from the linearity of the Schrödinger equation, the main equation of quantum theory, which describes the temporal evolution of an object called the *wave function*. To eliminate these unobserved macroscopic superpositions of incompatible states of affairs, standard quantum mechanics (as typically presented in physics textbooks) invokes the *collapse rule*. According to this rule, when a measurement is performed, the wave function no longer evolves according to the Schrödinger equation but instead randomly collapses into one of the terms in the superposition. After the measurement, the particle acquires a definite property, for instance being located "here."

The addition of this second evolution rule has been regarded by many realists as an unsatisfactory attempt to solve the problem, primarily because it is vague what qualifies as a measurement process. Nonetheless, structuralists have argued that their approach remains viable even within the quantum domain, provided that measurement apparatuses are not understood as being composed of microscopic particles, but rather as suitably emergent and useful patterns. I present structuralism, its motivations, and the objections to it in Chapter 2, after first outlining the basics of standard quantum mechanics. Although structuralism is a realist proposal, it remains open to various criticisms. Beyond internal objections, I argue that the kind of realism it promotes is too close to instrumentalism, as it emphasizes the usefulness of patterns and explains the data through *principles* that constrain phenomena, rather than through a constructive account based on the motion of microscopic components.

Guided by the need to provide a more precise theory and a more robust form of explanation, some have argued that for quantum theory to be compatible with realism, *the measurement problem must be solved precisely*, without relying on vague notions such as "measurement," as is done in structuralist approaches. Several solutions have been proposed along these lines, the most promising of which are the pilot-wave theory, the spontaneous localization theory, and the many-worlds theory. In Chapter 3, I present a view known as *wave function realism*, which some have suggested offers the best interpretation of these theories. According to this view, the wave function should be understood as a real, material field. A crucial feature of this approach is that this field exists in a high-dimensional space, and it has therefore been proposed that three-dimensional space, along with the macroscopic objects and properties we observe, should be regarded as emergent from this higher-dimensional framework. Both structuralism and wave function realism represent a radical departure from the pre-quantum understanding of how physical theories explain phenomena: the former challenges the existence of objects, while the latter questions the three-dimensionality of space itself. In particular, the entire realist enterprise must be reimagined: explaining a phenomenon no longer involves analyzing the behavior of its constituents from the bottom up, as in classical physics. Instead, new top-down explanatory strategies must be developed, either beginning with structure alone at the fundamental level, or with a material field in high-dimensional space.

Structuralists and wave function realists argue, albeit in very different ways, that *this is the cost* of maintaining realism in the quantum domain. *I disagree.* In the final four chapters of this book, I present, explore, and advocate for what I take to be the best approach to the metaphysics of quantum theory. My view is introduced in Chapter 4. While it is commonly referred to in the literature as the primitive ontology approach, I will try to avoid this term as much as possible. I have come to realize that the name makes the view seem more bizarre than it truly is. In fact, the opposite is the case: it is *the most obvious, commonsensical, simple, straightforward, symmetry-preserving and constructively explanatory approach to the metaphysics of quantum theory one can conceive.* The core idea is to seek a theory that offers the simplest constructive explanation of the phenomena, and that this description is as objective as possible: an approach that naturally extends the explanatory strategies that were successfully employed in physics before the advent of quantum theory. This approach requires assumptions that have served as the foundation of fundamental physics since Newton (if not earlier): namely, that matter is composed of microscopic three-dimensional entities evolving in time within a uniform universe. That is, the fundamental ontology of matter is *microscopic and spatiotemporal.* The three-dimensional nature of the material ontology, combined with the assumption that fundamental objects are sufficiently microscopic to constitute macroscopic objects, ensures a constructive explanation. As expected, the choice of the fundamental ontology of matter is not determined by the data. However, I argue that, given the theories we currently have, the most compelling choice is particles. This is because particles provide the simplest and most explanatory ontology while respecting symmetries. On another note, the laws governing the behavior of matter do not privilege any particular location in space or moment in time. Instead, the universe is uniform such that, *ceteris paribus*, the evolution of these entities proceeds in the same way everywhere. Mathematically, this is reflected in the *symmetries, or invariance* properties, of the theory. This requirement is often formulated in terms of the principle of relativity, which holds that the laws of physics must be the same everywhere, regardless of the perspective from which they are described, a guiding principle across many physical theories, from classical mechanics to the theory of relativity itself. Thus, I argue, not all spatiotemporal ontologies should be considered fundamental. For instance, in classical electrodynamics, where both particles and fields exist in three-dimensional space, we should consider only particles as fundamental. If we do otherwise, the theory would lose symmetries. Thus, the fundamental ontology of matter should consist of entities that are spatiotemporal and *foundational*—those that preserve symmetries. An appropriate term for my view might be the *spatiotemporal foundational ontology approach* (though admittedly, it's a mouthful). The term "spatiotemporal" grounds constructive explanation, while "foundational" ensures respect for symmetries. A theory that fits this description is the *de Broglie-Bohm pilot-wave theory* (or pilot-wave theory, for short), also known as Bohmian mechanics, which is fundamentally a theory of particles. This theory is compatible with a robust form of realism, as it explains phenomena constructively. Additionally, it preserves symmetries, which are crucial for making the theory perspective-free. However, the theory is *manifestly nonlocal*, meaning that instantaneous interactions between arbitrarily distant particles are possible. This is consistent with the violation of Bell's inequality, but it presents a significant challenge, as this nonlocality is in tension with relativity. Even though Lorentz-invariant pilot-wave theories have been proposed, they all contain a preferred frame (a preferred foliation of spacetime), which is considered to contradict the "spirit" of relativity. Interestingly, some foliation-free relativistic quantum theories have been proposed, but these no longer feature deterministic evolution. In the final section of Chapter 4, I discuss whether the tension between quantum theory and relativity can be mitigated by altering the theory's temporal evolution or by switching to a different fundamental ontology of matter. However, I argue that these alternative approaches do not justify their costs.

That said, while Chapter 4 focuses on matter, Chapter 5 shifts to discussing how the wave function should be understood within this framework. In a nutshell, my view is that the wave function characterizes the interaction of matter, rather than matter itself. I call this view *wave-functionalism*. According to this perspective, the wave function is an objective, nonepistemic, and nonrepresentational entity. My approach is a nonreductive functional one, where the wave function is defined by the roles it plays in the theory. In simple terms, *the wave function is as the wave function does*. I examine the empirical, explanatory, and physical roles of the wave function, which together allow it to effectively reproduce the phenomena in a constructive manner. Thus, the wave function does not represent material or nomological facts, but it is still objectively not epistemic. The central idea is that the wave function can be seen as a possible realizer of the functions it performs in the theory. This approach is *nonreductive* because the wave function is not eliminated; the information it contains must be included in the laws, alongside the distribution of matter, to form a complete theory. In the rest of the chapter, I compare my approach with other perspectives that treat the wave function as nonmaterial. One such perspective is the nomological approach, which views the wave function as a law of nature. The primary objection to this view—namely, that the wave function doesn't resemble a law because it is time-dependent and nonunique—is avoided in my approach, where the wave function is seen as an ingredient of the law rather than the law itself. Another alternative is the epistemic approach, which holds that the wave function reflects our ignorance of the system being studied. The main drawback of this view, namely its inability to account for objective phenomena like interference, is sidestepped in my approach, since the wave function is not material but is nonetheless objective. Finally, pragmatist approaches share with my view the idea that the wave function is defined by its roles in the theory. However, they equate these roles with prescribing the experimenter's expectations, offering a principle-based explanation rather than a constructive one, as I do. Therefore, *if someone is interested in a constructive explanation, I argue, they should endorse my view*. In the second half of Chapter 5, I explore the type of realism suggested by my proposal. Within the framework of *explanationism* (discussed in Chapter 1), I argue that the *wave function* should be understood as a *presuppositional posit*. In other words, the wave function is a provisional assumption that helps make the theory work, but we are not justified in being realist about it. This stance allows us to overcome the pessimistic meta-induction challenge, provided the fundamental ontology of matter in quantum theory is particles, as it is in classical physics.

In Chapter 6, I propose a potentially radical expansion of this approach: the idea that *fundamental objects have no intrinsic properties* beyond those needed to specify their nature. For example, particles are identified by their spatial location, but they do not inherently possess properties like mass or charge. In other words, particles are "bare" or "thin," and because of this, *they are all identical*. The motivation for this view arises from how my approach eliminates the contextuality of spin. As discussed in Chapter 4, if spin were considered a fundamental property of quantum objects, it would be contextual, meaning its value would depend on how it is measured. I believe the best way to understand this is to recognize that spin is not a property at all. This raises the question: What happens if we extend this idea to all properties? What if we "move" all properties, so to speak, "into" the law? This contrasts sharply with the traditional view, where the laws act differently on different objects because of their properties. For example, in a magnetic field, an electron will move up while a positron (a positively charged electron) moves down, due to their negative and positive charges, respectively. In contrast, my view holds that all entities in the fundamental ontology of matter are identical: they have no properties. Using the traditional metaphor of God creating particles, consider a theory of particles in which God creates all particles as identical and "bare." However, God also creates a law that acts on them differently, as if there were two distinct families of particles, such as electrons and positrons. Essentially, some particles are governed by a law that makes them move up in a magnetic field, and these are what we call electrons. The rest are governed by a law that makes them move down, and they are called positrons. To put it another way, being an electron means being a particle governed by an "up" effective law, and being a positron means being a particle governed by a "down" effective law; it is not about having a specific mass or charge. In this approach intrinsic properties are not entirely eliminated; rather, they are seen as nonfundamentally emergent, similarly as macroscopic properties such as solidity or solubility. Moreover, effective laws are best understood as the structure connecting the "thin" objects. Therefore, my approach can be directly compared to structural realism. While the traditional view can be seen as an object-oriented metaphysics grounded in properties, my approach could be described as a *thin-object-oriented metaphysics grounded in structure* (in the sense of law-like relations). Eliminativism, on the other hand, could be seen as a no-object-oriented metaphysics also grounded in structure. I argue that while my approach shares many motivations with structural realism—such as its simplicity and the fact that it does not need to explain what properties are or how they function—it avoids some of the major shortcomings of structuralism. This is because, in my view, there are still objects, even if they are thin. Moreover, my approach offers several advantages over the traditional view: it is more parsimonious by eliminating properties at the fundamental level; more fruitful because it extends naturally to the quantum domain; and more explanatory, as it resolves certain foundational problems.

I then address potential objections—that my proposal is too radical, mysterious, or even conspiratorial—and reply that it is necessary precisely because traditional metaphysics is itself contentious, that it is merely unfamiliar rather than arcane, and that it is no more conspiratorial than other frameworks.

In Chapter 7, I take a step back to reflect on why we find ourselves in a situation where interpretative freedom has led to an overwhelming number of alternative possibilities. I speculate on *what might have happened if history had developed more linearly*, and perhaps more rationally, than it did. I argue that it seems plausible that, had people attempted to systematically extend the successful constructive explanatory framework to the quantum domain, they would have naturally arrived at the pilot-wave theory. The fact that realists now think that making quantum theory compatible with a realist interpretation must involve *solving the measurement problem is merely a contingent effect* of how history played out. Since it focuses on dealing with unobservable macroscopic superpositions, rather than individuating a precise microscopic ontology with which to constructively explain the phenomena, the measurement problem is a red herring—a wild goose chase. Solving the measurement problem does not guarantee that the resulting theory will be constructive. And among the solutions to the measurement problem, *only the pilot-wave theory straightforwardly guarantees that.* Avoiding obsession with the measurement problem, and instead focusing on finding a constructive quantum theory, would have led to dismissing approaches involving emergence, structures, consciousness, many-worlds, or even stochastic theories.

Finally, I conclude the book with an epilogue, in which I revisit the key elements of my proposal and offer some perspectives for future work.

Audience

This book is aimed at philosophers, as well as physicists interested in the foundations of physical theories. Additionally, since I have made an effort to explain the technical terms I use, a reader without specialized knowledge but with an interest in these disciplines should still be able to follow along.

Formulas and Technical Language

I have made every effort to avoid unnecessary mathematical formalism: not because I believe there is anything wrong with using mathematics, but because I think it can hinder understanding if not essential for making the point. That said, I do occasionally introduce formulas to demystify the use of mathematics and to connect the discussion with relevant, more technical literature. Additionally, to clarify concepts that could reasonably be skipped, I use parenthetical phrases instead of endnotes, which I find distracting, as footnotes were not an option.

References

This is not a textbook, although I do not assume much prior knowledge. Rather, it is a research book that explores scientific realism in the quantum domain from a distinctive perspective. Accordingly, I do not cover all possible positions. Additionally, I do not provide extensive references within the text, as I believe they can distract from the main argument. However, at the end of each chapter, I list further readings that discuss issues I had to leave out of the main text, especially in the first two introductory chapters.

1

Scientific Realism and the Classical World

Roughly speaking, *scientific realism* is the view that scientific theories can faithfully and accurately inform us about the nature of the world. Since *metaphysics* is the study of the nature of reality, scientific realists believe that our best scientific theories can guide us in doing metaphysics. There are many interesting discussions related to scientific realism, such as its exact formulation, its connection to explanation and understanding, and its implications for the progress and goals of science. I will not address these issues here, not because they are unimportant or uninteresting, but because doing so would lead to a very different book. The book I want to write focuses on what I believe to be the best way to approach scientific realism in the context of quantum theory in the nonrelativistic regime. As such, I will gloss over many nuances and make some imprecise statements. While I will briefly review the main idea behind scientific realism (or "realism" for short), its key challenges, and some proposed solutions, this chapter is not meant to be a comprehensive treatise on realism. Instead, the discussion in this chapter is specifically tailored to the material presented in the following chapters, where I examine how quantum mechanics has raised new challenges for realism and how realists have responded to them. Please refer to the last section ("Further Readings and Discussion") for a list of interesting papers on the issues discussed in this chapter.

Chapter Schema Details

The first section presents the main argument in support of realism, the "*no-miracle*" *argument*. Its basic idea is that the best explanation for a theory's success is its truth. Assuming realism, I then present the picture of the world provided by classical mechanics: macroscopic objects are made of point-like particles in three-dimensional space, moving through time as described by Newton's laws. I also discuss classical electrodynamics as a theory of particles. In presenting these theories, I will explain some mathematical details that will be useful later.

In the second section, I address objections to realism based on general considerations about scientific methodology, specifically the "*pessimistic meta-induction*" argument and the problem of underdetermination. Roughly speaking, the former challenges the no-miracle argument by asserting that a theory's success is not evidence of its truth: in the past, many successful theories have turned out to be false. The problem of *underdetermination*, on the other hand, is the more general issue that there is an infinite number of incompatible theories that can account for the same data, meaning experiments cannot definitively tell us which theory is true.

In the third section, I summarize *some responses* to these challenges. I survey the key components of *structuralism and explanationism*—two responses to the pessimistic meta-induction argument, which I explore in more detail in later chapters (structuralism in Chapter 2 and explanationism in the second half of Chapter 5). Both approaches share the idea of restricting realism to, respectively, the theory's structure and the parts of the theory that play an explanatory role. If one assumes that only part of a theory is true, rather than the entire theory, and if one can show that the true part is preserved during theory change, then one can counter the pessimistic meta-induction argument. Regarding the underdetermination challenge, I discuss the traditional response of invoking "*extra-empirical*" *virtues*: features like simplicity or explanatory power that can guide theory selection. I introduce a new perspective by discussing the explanatory schema of classical mechanics in terms of the *distinction between principle and constructive theories*. I call constructive theories the ones which understand macroscopic objects as composed of microscopic fundamental entities, with macroscopic laws being derived (under suitable approximations) from the microscopic dynamics. In this sense, they are reductionist, bottom-up approaches. In contrast, principle theories offer top-down perspectives, where phenomena are systematized by principles and where microscopic dynamics or compositionality do not play a significant role. I conclude by discussing *the meaning and role of symmetries* in theory evaluation. I argue that symmetries are features that a theory should possess to make our description of the world as perspective-free as possible. Ultimately, I will argue that constructive explanations that preserve symmetries should be preferred, as they provide a deeper understanding.

The No-Miracle Argument and Classical Mechanics

Metaphysics can be characterized in many ways, but the definition I am concerned with pertains to the study of the nature of the world. Ontology, in particular, addresses questions such as what entities exist in the world. Using a common metaphor, what did God need to create in order to bring about a world like ours? In the philosophy of science, scientific realism generally maintains that our best scientific theories can reliably inform us about the nature of the world. For instance, if our best fundamental physical theory describes point-like particles, then we are justified in believing that such objects, with no internal structure, really exist. These point-like particles, therefore, form the ontology of the theory. There is an obvious sense in which *metaphysicians and scientific realists are trying to answer the same questions*. Because of this, I believe that their respective disciplines should inform one another. Indeed, doing metaphysics by looking at the sciences is often referred to as naturalized metaphysics, in contrast to armchair metaphysics, where one hypothesizes about the nature of reality based on our concepts. I will not attempt to criticize the latter or positively argue for the former, as that would deviate too much from the focus of this book. Moreover, even if metaphysics can be naturalized, it does not necessarily follow that this is the only legitimate way of doing metaphysics.

I want to emphasize that I will restrict the discussion to realism in connection with our *fundamental physical theories*. While arguments for realism can be made in the context of other types of theories (e.g., biological theories), I will not be addressing them here. A fundamental physical theory is one that describes the nature and behavior of physical entities that are considered fundamental. These are entities without internal structure, deemed indispensable and primitively postulated to exist as the basic components of everything in the universe. To use the earlier metaphor, these are the entities that God had to create to form the material constitution of the universe as we observe it. These fundamental entities represent the material ontology of the world—what tables and chairs, for example, are fundamentally made of. Often, in our theories, fundamental entities are microscopic: in classical (Newtonian) mechanics, they are point particles that are not observable to the naked eye. The term "microscopic" is understood in contrast to "macroscopic," which refers to objects that are part of our everyday experiences. In addition to fundamental physical theories, there are other physical theories that are not considered fundamental. For example, hydrodynamics is a physical theory but not fundamental, as it describes the behavior of fluids (liquids and gases), which are taken to be composed of Newtonian particles. Nonfundamental theories are often phenomenological, meaning they describe laws as regularities in macroscopic phenomena, establishing principles of what can or cannot happen without explaining the underlying reason. In the terminology I introduce later in this chapter, these are principle theories, where the explanation of a macroscopic phenomenon is not based on its microscopic constituents but on unifying principles. In contrast, many, including myself, believe that a theory must be a dynamical, constructive theory to be fundamental: it must explain phenomena in terms of the dynamics of its ontology. Current candidates for fundamental physical theories, understood in this way, are quantum mechanics and relativity theories. I will not discuss relativity in detail, as I focus on developing a metaphysics for quantum mechanics. While some have argued that quantum mechanics is incompatible with realism (as discussed in the next chapter), it is now generally accepted that this is not true. In this book, particularly in Chapter 4 and the following chapters, I will argue for a way of doing naturalized metaphysics from quantum mechanics that respects a robust form of realism, where phenomena are explained compositionally and dynamically, preserving symmetries. Given these qualifications, it should be clear to the reader that, when I use the term "theory" for simplicity, I am specifically referring to a "fundamental physical theory."

The Ultimate Argument for Realism

Arguably, the main reason to believe in realism stems from the observation that the only alternative explanation for a theory's empirical success seems absurd. This intuition lies at the heart of the so-called ultimate argument, or *no-miracle argument* (Putnam 1975, Boyd 1980, Musgrave 1988). If a theory is empirically successful, meaning it satisfactorily reproduces experimental data and predicts new phenomena, the natural question is: Where does its success come from? The no-miracle argument asserts that *the success of a theory should be taken as evidence of its (approximate) truth.* (Again, I assume that the notion of truth and approximate truth is understandable, setting aside debates over their exact nature.)

The idea is that the best explanation for a theory's success is that it truthfully describes the world. Consider a theory that can correctly predict and reproduce macroscopic appearances based on the motion of microscopic point particles. The realist argues that the best explanation for this empirical success is that such particles exist in the world. After all, what is the alternative? It is conceivable that the theory could be successful without the assumption that what it postulates—point particles—actually exist. However, if that were the case, the theory's success would be the result of a cosmic coincidence, a massive conspiracy, a global deception—a miracle: everything would align so perfectly that it appears macroscopic objects are made of point particles, even if this is not actually true.

So, either the world is truly made of particles, as the theory suggests, or the theory is false and some miracle occurs, with everything conspiring to make it seem as though the world is made of particles, despite the absence of such particles. Realists argue that the latter option is implausible, and, assuming that we are justified in believing what best explains the phenomena, they conclude that we are justified in believing the theory is true because of its success.

An Example of Naturalized Metaphysics

There are powerful arguments against this position, as discussed in the next section. However, before diving into those, let me emphasize that realism has been the starting hypothesis for many successful fundamental physical theories in the past. For example, *Newton was a realist*: he used arguments based on realism to counter Leibniz's view of space. Leibniz believed that space was nothing more than a collection of relations between material bodies. In contrast, Newton argued that there is something beyond these relations—namely, absolute space—and that it provides the best explanation for phenomena like the body's acceleration (the so-called inertial effects). For example, the concave shape of the water in a spinning bucket is evidence for the acceleration of the water with respect to absolute space.

Assuming realism about space (and time), a theory must also clearly specify what exists and constitutes material objects that reside in space and evolve over time. In the case of classical mechanics, *the metaphysical hypothesis is that matter is made up of particles without internal structure*: these particles form the material ontology of the theory. Since physics uses mathematics, Newton's space is represented as a three-dimensional Euclidean space $\mathbb{R}^3$, and particles are represented as points in that space, with their position written as $r \in \mathbb{R}^3$. These particles also have mass, represented mathematically by the positive real number m, and change their position over time, represented by the positive real number t. In other words, each particle has a trajectory, expressed by the function $r(t)$, which specifies the different spatial locations occupied by the particle at different times. Newton's theory allows one to compute the trajectories of particles as solutions to Newton's equation, which describes the temporal evolution of their position: $r = r(t)$. A solution to this equation represents a possible state of the world according to the theory. In mathematical terms, Newton's equation is $F = ma$, which states that the acceleration a of an object with mass m—the rate of change of its velocity v—changes over time due to the presence of forces F. In simpler terms, forces are responsible for a body's motion. For clarity, let me define a few terms. The velocity is the rate of change of the particle's position, and mathematically, it is the derivative of position with respect to time: $v(t) = \frac{d}{dt}r(t)$. (Sometimes the derivative with respect to time is written with a dot on top of the variable, as $v = \dot{r} \stackrel{\text{def}}{=} \frac{d}{dt}r(t)$. The derivative of a general function $f(r)$ with respect to time is written as $f'(r) \stackrel{\text{def}}{=} \frac{d}{dt}f(r)$, or its gradient $f'(r) \stackrel{\text{def}}{=} grad\, f \stackrel{\text{def}}{=} \nabla_r f(r)$.) The acceleration, being the derivative of velocity, is therefore the second derivative of position: $a(t) = \frac{d}{dt}v(t) = \frac{d^2}{dt^2}r(t)$, which is sometimes written as $a = \ddot{r}$. This makes Newton's equation a second-order differential equation for r(t) $m\frac{d^2}{dt^2}r(t) = F$. Notice that position, velocity, acceleration, and force are vectors. Geometrically, a vector is an arrow with both direction and magnitude. In contrast to a scalar, which is represented by a single number, a vector in three-dimensional space (such as Newton's space) can be represented by an ordered triplet of numbers. These represent the vector's coordinates in a given frame of reference or coordinate system. The same arrow can be represented by different sets of numbers in different coordinate systems, meaning a vector is not uniquely identified by a single ordered set of numbers—it represents the same object from different perspectives. Therefore, Newton's equation is a vector equation. Since this is not a formal textbook, I will not use the common notation for vectors, which involves either placing a little arrow on top of the variable (e.g., $\vec{r}$) or using boldface (e.g., $\boldsymbol{r}$). If the vectorial nature of an object is important, I will make sure to note it explicitly.

If we only consider particles with mass, as Newton did, the force acting on the particles is the gravitational force F_g. This force is generated by a body with mass M, which affects another body with mass m in a way that is directly proportional to the product of the masses and inversely proportional to the square of their relative distance. Mathematically, introducing the constant $G = 6.6743 \times 10^{-11} m^3\, kg^{-1}\, s^{-2}$ to define the strength of the gravitational interaction, we write: $F_g = G\frac{mM}{r^2}$. (Note that the force is a vector, directed along the line connecting the two objects.) Sometimes, it is convenient to express the force in terms of another function, called the potential, usually denoted by V, which is defined such that the force is the negative derivative of the potential with respect to position: $F = -\frac{d}{dr}V = -V'$. In this case, the potential is given by $V = -G\frac{mM}{r}$. The potential V is a scalar function and is defined up to a constant. (That is, the potential V and the potential $V + c$, where c is any constant, give rise to the same force because the derivative of a constant is zero. Therefore, the two potentials result in the same particle trajectory.)

A Different Possible Ontology

Newton assumed that *his theory was complete*: he wanted it to apply to everything, so that all phenomena at all scales could, in principle, be understood in terms of its fundamental particle ontology. In fact, Newton even explored the possibility that *light is made of particles*, although he was aware that he did not have a fully developed theory in this regard. Geometric optics, on the other hand, is a phenomenological theory that describes the laws governing the behavior of light, including its rectilinear propagation and the laws of reflection and refraction, which explain what happens when light encounters different surfaces. If Newton's theory is complete, then the laws of geometric optics should be derivable from it. To some extent, this is the case: like all particles when they are not subject to forces, light particles travel in straight lines. Also, when encountering hard surfaces, light particles rebound in accordance with the law of reflection. When traveling from air to water (for example), the light particles bend due to water's greater gravitational pull, which is consistent with the law of refraction. However, other experiments show that light also displays phenomena like interference and diffraction, which are more difficult to explain with a particle theory. Consider the case of a series of light particles directed toward a slit in a barrier, and then traveling toward a detection screen. A particle theory would predict that some light particles will hit the barrier and bounce back, while others will pass through the slit and hit the screen, leaving a spot of the same size and shape as the slit. Instead, the observed spot is larger in shape, indicating that light spreads around obstacles, unlike particles. This phenomenon is called *diffraction*. Similarly, if light were made of particles, one would expect to observe two spots on the screen corresponding to the two slits in an experiment where light is directed at a barrier with two slits. However, the two-slit experiment produces a complex and periodic distribution of shadowed regions, as if the beams from the two slits were *interfering* with each other. In principle, a particle-based explanation could still account for these phenomena, but it would require *ad hoc* modifications, such as assuming that the particles have different shapes.

An alternative theory, proposed by Huygens, Young, Fresnel, and others, suggested that *light is a wave*. In general, a wave is an oscillation—an up-and-down or back-and-forth motion of particles within a substance, which is called the medium. A typical example of a wave is sound: it involves the oscillation of particles in the medium, propagating at a finite velocity. According to the wave theory, light is a longitudinal wave: an up-and-down oscillation of particles in a medium known as the ether. (Although the ether theory was later abandoned, requiring it to have very unusual properties, numerous experiments aimed at detecting it were conducted and ultimately considered unsuccessful. This aspect of the theory is worth mentioning for historical context.) The wave theory of light could naturally explain phenomena like interference and diffraction, which the particle theory struggled to account for, as these are properties characteristic of waves. To understand this, imagine throwing a pebble into a pond: it creates waves in the water. When a wave passes through a single slit, it spreads out over a larger area than just the slit, which is called diffraction. If multiple stones are thrown into the water, interference occurs: at certain points, the waves combine to increase the amplitude (constructive interference), while at others, they cancel each other out (destructive interference). Similarly, interference can be observed with two openings (or slits) in a barrier in the water: the waves coming from each slit combine and interfere with each other. Additionally, the wave theory of light can also account for the laws of geometrical optics. Light emitted from a point source spreads out in all directions at a finite velocity, forming wavefronts—a set of particles oscillating together—which take the shape of a growing sphere over time. When the source is far enough away, the wavefronts can be approximated as flat planes. Every time the oscillation reaches a new particle, it transfers its motion to that particle, generating a new wave. This is *the Huygens-Fresnel principle*, and it can explain phenomena like reflection and refraction: each particle on the reflecting (or refracting) surface generates a new wave, and the sum of these waves gives rise to the reflected (or refracted) wave, with a velocity that depends on the medium. Similarly, diffraction is explained by the generation of new waves.

Let me briefly explain *how waves are mathematically described*, as the formalism of the wave theory of light is quite similar to that of quantum theory. Additionally, this discussion will highlight that the reason waves interfere and diffract lies in the linearity of their evolution equation, which will be important when we discuss quantum theory. Typically, a wave (at least the kind of wave light is often modeled as) is described by a function u that varies with space and time. This function can be split into two components. Let me briefly explain how waves are mathematically described, as the formalism of the wave theory of light is quite similar to that of quantum theory. Additionally, this discussion will highlight that the reason waves interfere and diffract lies in the linearity of their evolution equation, which will be important when we discuss quantum theory.

Typically, a wave (at least the kind of wave light is often modeled as) is described by a function u that varies with space and time. This function can be split into two components, F and $G: u(r,t) = F(r - vt) + G(r + vt)$. Here, v is the velocity of propagation, r is the direction of propagation, and t is time. The function F represents a wave traveling forward, while G represents a wave traveling backward in time. This expression is the solution to an evolution equation, such as $\frac{1}{v^2}\frac{\partial^2}{\partial t^2}u(r,t) = \frac{\partial^2}{\partial r^2}u(r,t)$ (Note: The "slanted" ∂ symbols indicate partial derivatives, used when dealing with functions of multiple variables, like $u(r,t)$. These derivatives are taken with respect to just one variable while holding the others constant. Other forms of this equation are: $\Delta u - \frac{1}{v^2}\ddot{u} = 0$ or $\Box u = 0$, where Δ (also written as $\nabla \cdot \nabla$ or ∇^2) is the Laplacian operator, and $\Box$ is the d'Alembertian operator, defined as $\Box \stackrel{\text{def}}{=} \Delta - \frac{1}{v^2}\frac{\partial^2}{\partial t^2}$.) Waves are periodic, meaning they repeat in both space and time. They can be characterized by their wavelength λ (the spatial distance over which the wave repeats) and their frequency f (the number of oscillations per unit time). The frequency is the inverse of the period T (the time it takes for the wave to repeat). The relationship between wavelength and frequency gives the velocity at which the wave propagates: $\lambda f = v$. Other quantities often used to describe waves include the wave number $k = 2\pi / \lambda$ (which represents the number of wavelengths per unit distance) and the angular frequency $w = 2\pi f$ (another way to measure how fast the wave changes). The simplest form of a wave can be expressed using trigonometric functions. For example: $u(r,t) = A\cos(kr - wt) + A\cos(kr + wt)$, where A is the amplitude of the wave (and for simplicity, we assume the amplitude is constant in time). Another common form for a wave is: $u(r,t) = Ae^{\pm i(kr - wt)}$, where $kr - wt$ is called the phase, usually denoted as φ.

Waves like the ones described so far have a single wavelength and are referred to as monochromatic waves, or plane waves, because their wavefronts are perpendicular to the direction of propagation. However, in general, waves are often combinations or superpositions of many monochromatic components. This is an example of the *superposition principle*: if u_1 is a solution to the wave equation and u_2 is another solution, then the sum $u_1 + u_2$ is also a solution to the wave equation. Calling this a "principle" is somewhat misleading, as it is simply a consequence of the mathematical property of linearity. A linear equation is one in which the highest exponent of the variables and their derivatives is 1. For example, Newton's equation is not linear in r (since the highest exponent is 2, due to the second derivative of r), which means that for particles, the superposition principle does not apply. This is exactly how it should be, since particles do not follow the same rules as waves. In fact, linearity is essential to account for wave phenomena such as interference and diffraction, which arise from the superposition principle. *Physically, the presence or absence of interference and diffraction signals different ontologies: waves versus particles.* A particle is a localized entity, so it doesn't make sense to think of it as "superimposing" with another particle. In contrast, a wave describes something that spreads over a larger region of space, allowing it to interact with another wave. Unlike particles, two waves can occupy the same spatial region simultaneously. This is the core idea behind the superposition principle (see Figure 1.1), which explains phenomena such as the spreading of waves around obstacles (known as diffraction) and the pattern observed in the two-slit experiment (known as interference).

Waves as Fields

After Newton, many new phenomena were discovered, including electricity. At the end of the nineteenth century, James Clerk Maxwell summarized the work of many preceding physicists by proposing a set of new laws. These laws introduced a new property of matter—charge—which is typically denoted by q and is mathematically represented by a real number. Initially, electricity was thought to be a fluid, but it was

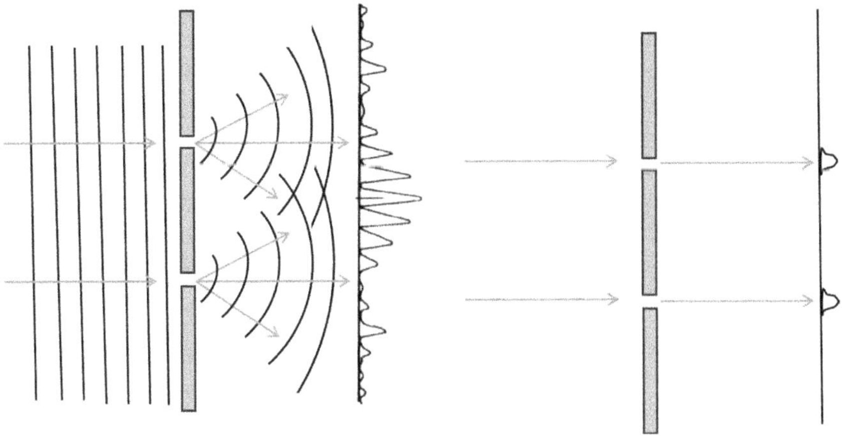

Figure 1.1 The superposition principle does not hold for particles.

later understood to consist of the motion of charged particles, known as electrons. Due to their charge, particles experience a new force, F_e, called the *electric force*. This force has a similar form to the gravitational force, in that it is directly proportional to the product of the charges and inversely proportional to the square of their relative distance. A particle with charge Q generates a force that affects another charged particle q according to the following equation: $F_e = k\frac{qQ}{r^2}$, where k is a constant, and r is the distance between the charges. The constant k is defined as $k = 1/4\pi\epsilon$, where ϵ is the permittivity of the medium, with $\epsilon = \epsilon_0\epsilon_r$. Here, ϵ_0 is the vacuum permittivity, a constant with the value $8.85 \times 10^{-12} s^4 A^2 kg^{-1} m^{-3}$ and ϵ_r is the relative permittivity that depends on the material.

Unlike the gravitational force, which is always attractive, the electric force can also be repulsive. This is because, unlike masses, there are two types of charge: charges of the same type repel each other, while charges of opposite types attract. By convention, these charges are labeled as positive and negative. Additionally, *the electric field* E can be introduced, which is defined as the force generated by a charge Q and experienced by a "test" particle with charge q. The field is calculated as the force per unit charge, so $E = F_e/q$. The force F_e acting on q due to Q can thus be written as $F_e = qE$. When a charged particle is in motion, another force called the Lorentz force, F_L, comes into play. This force involves a new field, B, known as the *magnetic field*. The Lorentz force is given by the equation $F_L = qv \times B$, where v is the velocity of the charged particle q, and the $\times$ symbol denotes the vector cross product. This results in a force that is perpendicular to both the velocity of the particle and the magnetic field.

Initially, fields were considered simply as *useful tools* for obtaining empirically adequate trajectories for particles. The idea was that particles, which constituted the fundamental building blocks of matter, had a new property—charge—that made them interact with other charged particles. The electric and magnetic fields conveniently summarized these interactions, but beyond that, they were not seen as having any independent existence. However, over time, people began to take more seriously the idea that these fields might actually exist "out there," alongside the particles. In other words, it was proposed that there are both particles—point-like entities—and electric and magnetic fields, which are nonlocalized, spread-out entities. Fields were now understood to be generated by charged particles, "expressing" the effects of one charge on another. These fields were shown to obey a set of equations, proposed by Maxwell, that are interconnected and can be combined into a wave equation describing the propagation of electromagnetic waves at the speed of light, $c: \frac{\partial^2 E}{\partial x^2} = \frac{1}{c^2}\frac{\partial^2 E}{\partial t^2}; \frac{\partial^2 B}{\partial x^2} = \frac{1}{c^2}\frac{\partial^2 B}{\partial t^2}$. This strongly suggested that *light should be understood as an oscillation of the electromagnetic fields propagating through space* (previously thought of as the ether), thus supporting the idea of light as a type of wave. While particles compose matter, it was now clear that electromagnetic fields, by oscillating, are responsible for light. This forms the basis of what is often called electromagnetic optics. It can also be shown that electromagnetic optics reduces to geometric optics under the "eikonal approximation." This approximation applies when the distances and time intervals are large compared to the wavelength and period of light. Under these conditions, the wave equation above reduces to a plane wave equation, from which the laws of geometrical optics can be derived. This reinforces the idea of light as an electromagnetic wave. Additionally, it was shown that these fields contribute to the total energy of the system, further supporting the notion that fields are part of the ontology of the theory, alongside particles.

Another key reason for introducing the concept of fields relates to the nature of forces in classical mechanics and electrodynamics as originally formulated. In these theories, forces act instantaneously, which contrasts with the idea that interactions between systems should propagate at a finite velocity. This notion is central to the principle of *local causality* (or locality), which will be discussed further in subsequent chapters. The instantaneous action of forces, no matter how distant, violates the idea that causes precede their effects. By introducing fields, this problem is resolved. Instead of thinking of forces as acting instantaneously at any distance, we can think of fields as mediators of the force. These fields can propagate through space, with interactions occurring at a finite speed. This view was further developed by Einstein and Infeld in 1938, helping to *shift our understanding of forces from being nonlocal (instantaneous) to being mediated through fields that propagate with finite velocity.*

Underdetermination and the Pessimistic Meta-Induction

Newtonian mechanics and classical electrodynamics propose *different ontologies*: Newtonian mechanics is built on particles and forces, while electrodynamics includes both particles and fields. Nonetheless, both theories exemplify a form of *naturalized metaphysics*: they aim to describe what the world is made of and how it behaves. In doing so, they both serve as compelling examples of the success of scientific realism, the idea that science aims to describe a mind-independent reality and often succeeds in doing so. From this perspective, realism can seem almost self-evident. Denying it raises difficult questions. *If realism is false, then how do we explain the success of science?* What are theories even for, if not to tell us what the world is like? These concerns become even more pressing when we move beyond physics. For example, medicine tells us that some illnesses are caused by viruses and bacteria. If realism is false, then we are not justified in believing that such entities exist. But then how do we explain why people get sick? Why do we sanitize surfaces or wash our hands, if not because we believe that doing so eliminates real, disease-causing organisms?

Despite these intuitive concerns, some philosophers argue that science need not aim to describe the world as it is. One influential alternative to realism is *instrumentalism*, the view that scientific theories are simply tools or instruments for organizing observations and making predictions. On this view, theories are not literally true or false; they are judged by their usefulness. A theory is good if it helps us predict experimental outcomes and manage our interaction with the world; if it tells us what to expect, not what is real. There are also more nuanced anti-realist positions. One of the most prominent is *constructive empiricism*, proposed by Bas van Fraassen (1980). According to this view, the goal of science is not truth, but empirical adequacy: theories should correctly describe what we can observe, but we need not commit to the existence of unobservable entities they postulate. Constructive empiricism and other modern forms of anti-realism thus advocate for a cautious, agnostic stance regarding the truth of our best theories, especially in relation to their claims about unobservable aspects of the world. But why adopt such a seemingly pessimistic view of science? *Why reject the idea that science gives us access to truth*? Anti-realists are often motivated by *powerful philosophical arguments against realism*, which aim to show that belief in the truth, or even approximate truth, of our best scientific theories is not justified. These arguments, which will be explored further in the following sections, challenge the realist's confidence in the ability of science to reveal the true nature of reality.

The Pessimistic Meta-Induction Argument

One powerful objection to scientific realism stems from the observation that many of our past scientific theories turned out to be false, so it seems likely that our current theories will eventually be proven false as well. In other words, *we appear to have inductive evidence against realism*: the history of science looks like a long sequence of failures, where every theory is eventually replaced. According to this line of thought, the fate of all theories is to be rejected. This view, sometimes called "historical pessimism," can be resisted through what's known as the "Animal Farm" move. Inspired by George Orwell's famous line—"All pigs are equal, but some pigs are more equal than others"—the idea here is: "*All theories are false, but some are more approximately true than others*" (Newton-Smith 1981). In this view, while no theory may be strictly true, some come closer to the truth than others. One way to defend this is by pointing to the greater predictive power of current theories compared to older ones. This leads to a variant of the no-miracle argument: the best explanation for a theory's success is that it is (approximately) true.

However, the *pessimistic meta-induction* challenges this reasoning. It argues that empirical success is not a reliable indicator of truth, not even approximate truth. Our current theories, despite their success, are likely to be false—just as all successful past theories eventually proved to be. Several historical examples illustrate this point. In astronomy, the once-successful crystalline sphere theory posited that celestial bodies moved within perfect concentric spheres rotating around the Earth. It was later replaced by more accurate astronomical models that entirely abandoned the notion of crystalline spheres. In ancient medicine, the humoral theory maintained that health depended on balancing bodily fluids, or "humors." A cold, for example, was believed to be caused by excess phlegm and treated with hot tea; a fever might be cooled with alcohol. Although this theory was once widely accepted, it was eventually replaced by modern medicine, which attributes illness to microorganisms and other factors, not imbalances in humors. Likewise, the phlogiston theory of combustion proposed that burning objects released a substance called phlogiston. This was later replaced by the oxygen theory, which correctly explains that combustion involves the intake of oxygen. Similarly, the caloric theory of heat treated heat as a material substance called "caloric," responsible for temperature differences. This theory, too, was abandoned in favor of the kinetic theory, which understands heat as molecular motion. Perhaps the most significant example is Newtonian mechanics. It was immensely successful: it accounted for data, made accurate predictions, and unified various phenomena such as planetary orbits and falling objects. Yet, Newtonian mechanics was eventually replaced by quantum mechanics and relativity, which revealed limitations in Newton's framework. In each of these cases, the theory was successful—and yet, false. We once believed in crystalline spheres, humors, phlogiston, caloric, and Newtonian particles because of the success of their respective theories. But we were wrong. Their success, in hindsight, was not good evidence of their truth.

The pessimistic meta-induction urges us to consider that there is no reason to believe quantum theory will fare any differently. Yes, it is currently successful, but so were many earlier theories, especially classical mechanics. It is likely that quantum theory will also be replaced. Indeed, there are well-known tensions between quantum mechanics and relativity (to be explored in Chapter 4), and many physicists are actively searching for new, more fundamental theories. Thus, despite their empirical success, we have no solid justification, contrary to the no-miracle argument, for believing that our current best theories, such as quantum mechanics, are true. The history of science cautions us against such confidence.

Underdetermination of Theory by Data

Another problem for realism is that any theory is always *underdetermined by the data*: there are invariably an infinite number of metaphysically incompatible but empirically indistinguishable theories. In other words, it is possible for theories to be mutually exclusive even if they are compatible with the same set of empirical phenomena. This means that *empirical evidence alone cannot tell us which theory is true*. Consider the case of an analog watch. How does it tell time? The answer seems easy: with its hands. However, these hands could be moved by either an electric motor or a mechanical one. Furthermore, each of these motors could be implemented in various ways. A mechanical motor, for example, could have a different number of springs, cogwheels, and so on. If we could open the watch, we could empirically discover how it functions. However, if we cannot open it, the various theories about how the watch works are underdetermined: there is no observation that can tell us the watch's inner workings.

Examples of underdetermination also arise in scientific theories. For instance, consider Newtonian mechanics applied to the entire universe. *It is underdetermined whether the universe is at absolute rest or moving with constant velocity* because the observable quantity is acceleration. We do not observe uniform motion, only accelerated motion. Accelerated bodies follow curved trajectories, while bodies in uniform motion and at rest both move in straight lines. (Zero or constant velocity gives rise to the same acceleration because acceleration is the derivative of velocity, and the derivative of a constant is zero.) So, how can we tell which is the true state of the universe if we cannot observationally distinguish between uniform motion and rest? Moreover, the same force (and thus the same particle trajectories) could be generated by two potentials that differ only by a constant (see previous section). If we take potential to represent something physically real, there are two theories—one for each potential—that are empirically indistinguishable. How can we tell which is correct? Perhaps even more interesting is the underdetermination between the *Newtonian, Lagrangian, and Hamiltonian formulations* of classical mechanics. The Newtonian approach, as discussed in the previous section, is one of the standard formulations. However, two other formulations are often presented in physics textbooks. One can rewrite Newton's law in terms of a new set of coordinates and a new function called the Lagrangian $(\mathcal{L})$, resulting in a new set of equations known as the Lagrangian equations, named after the physicist Lagrange. Alternatively, one can write the equations in terms of a different set of coordinates, using the Hamiltonian function $(\mathcal{H})$, which leads to the Hamiltonian equations, developed by Hamilton. Since these formulations are essentially re-expressions of Newton's equations in different variables, they do not lead to observable differences. Most physics students are taught that these reformulations are simply ways to simplify solving certain problems, and that the "true" picture of the world is provided by the Newtonian formulation. However, some have argued that these other frameworks may also offer valuable insights into reality (see "Further Readings" section). If so, how can we determine which formulation provides the best description of reality? Another example is given by *Poincaré's disk*, a world in the shape of a disk inhabited by two-dimensional creatures. The disk has a hot center, with decreasing temperatures as one moves toward the edges. Everything, including the creatures and their measuring devices, shrinks with temperature. The creatures will never be able to reach the edges, because as they move, they shrink. As a result, the world appears infinite to them. Additionally, the sum of the internal angles of a triangle will be less than two right angles due to the shrinking forces. There are at least two ways to interpret this evidence. One option is to assume that space is Euclidean and postulate the existence of a force that distorts measuring rods and everything else. Alternatively, if space is non-Euclidean, one could explain the phenomenon without invoking any forces. In certain non-Euclidean spaces, the internal angles of a triangle are indeed less than two right angles, so the creatures would measure this regardless of any shrinking forces. Which of these incompatible alternatives is correct? One cannot determine this through empirical observation alone: the two theories—the one that suggests a distorting force in a Euclidean space, and the one that posits a non-Euclidean world without such forces—are underdetermined by data.

Notice that *the empirical inadequacy of a theory can always be bypassed*, one way or another. In fact, one might think that we can rule out theories that are not empirically adequate by performing the right type of experiments. This is the core idea behind falsificationism: one can refute a theory if it does not correctly reproduce the data (Popper 1935). However, this view is naïve. When making experimental predictions, we assume more than just the theory we want to test. We also rely on many (often unspoken) assumptions that we take for granted. For instance, when observing the orbit of a distant planet, we assume the truth of the laws of optics that make the telescope a reliable instrument (Duhem 1906). So, in principle, we can counter falsifying evidence for our theory by blaming some auxiliary hypothesis—this might even include the laws of geometry (as in Poincaré's example above) or the laws of logic, as famously suggested by Quine (1951). This *ambiguity in falsificationism is known as the Quine-Duhem thesis*: one can hold on to their preferred theory no matter what, by suitably modifying other assumptions. Notice that this remains true even if the theory is initially not empirically adequate: one can make a theory empirically adequate by adjusting something else in our set of assumptions (the "web of beliefs"). However, while most would agree on which strategies to avoid falsification are reasonable and which are not, there are still differences in judgment. For example, the Ptolemaic objection to Galileo's use of the telescope to save their theory is generally regarded as unreasonable, while Copernicans questioning the assumption that there are seven planets, rather than questioning their theory, seems more justified. Nonetheless, I will argue (in Chapter 7) that this is not what happened in the history of quantum mechanics. Many metaphysically extravagant theories, such as those proposing the existence of an infinite number of unobservable worlds or that consciousness plays an active role in physics, would not have been considered under different circumstances.

That said, even granting that we can rule out empirically inadequate theories, underdetermination creates a problem for realism. This is because we do not have the empirical means to decide which theory is true. If there are no other means of adjudicating between theories other than observation—and all of them are compatible with the same data—then we are not justified in picking one over another. As a result, we are not justified in believing which of the two incompatible alternatives should guide our metaphysical understanding.

Restricting Realism and Super-Empirical Virtues

There are many ways to *defend realism* against these charges, but for the sake of brevity and relevance, I will only review a few. In the following sections, I briefly discuss some responses to the pessimistic meta-induction argument and the challenge of underdetermination.

Restricting Realism

Some have responded to the pessimistic meta-induction argument by claiming it commits the *base rate fallacy*: there are many more false successful past theories than true ones. Therefore, the fact that past theories turned out to be false merely suggests that truth is rare, and this doesn't say anything about the relationship between truth and success (see, notably, Lewis 2001). However, the pessimistic meta-induction argument can also be approached *deductively*: even a single successful but false past theory undermines the explanatory connection between success and truth. In response, one could argue that *success is explained by truth, even if it does not conclusively prove the theory's truth*. In other words, successful theories can be false without serving as counterexamples to realism. Others have pointed out that the number of empirically successful-but-false theories is dramatically reduced if we consider only sufficiently mature and *non-ad hoc theories*. However, it is neither easy nor uncontroversial to define what qualifies as a "mature" or "non-*ad hoc*" theory. Alternatively, some have proposed approaches to "*restrict*" *realism*. That is, one might argue that we are entitled to be realists about a particular set of entities posited by the theory, rather than insisting on realism about the entire theory. The idea here is to focus on what it means for a theory to be empirically successful, and then identify the aspects of the theory responsible for that success. Imagine that a theory is empirically successful because it is both explanatory and predictive. Suppose, one day, that theory turns out to be false and is replaced by another, even more successful theory. If we can show that the entities responsible for the original theory's success are retained in the successor theory, then the pessimistic meta-induction argument is undermined. Interpreting the no-miracle argument as suggesting that the success of a theory provides evidence for the existence of the entities that make it successful, we could still maintain that we are justified in believing these entities exist, even after theory change. By restricting realism, we offer an alternative explanation for the success of past false-but-successful theories: they were successful not because the entire theory was true, but because some parts of it were true.

There are various ways to restrict realism, and several are particularly relevant to this book. One such approach, known in the literature as the *divide et impera* strategy (Kitcher 1993, Psillos 1999), is what I prefer to call *explanationism*. This view proposes that one can be a realist about the objects in a theory that are responsible for the theory's explanatory power. Explanationists distinguish between "working" and "presuppositional" posits of a theory. The *working posits* are those responsible for the theory's success, while the presuppositional posits are theoretical constituents necessary to complete the theory. According to this view, if the working posits are preserved during theory change, it suggests that past theories were successful because they got the working posits right. At the same time, these theories are false in their entirety because they got the presuppositional posits wrong. Thus, realists are justified in believing in the entities represented by the working posits, but they are not obligated to commit to the existence of what the *presuppositional posits* describe since these are merely "idle" components that make no difference to the theory's success. To see how this approach works, consider the wave theory of light and its successor, Maxwell's electrodynamics. The wave theory succeeded because it got the working posit right—the idea that light is a wave. Its nature as a wave is responsible for the theory's success, and it is preserved in electrodynamics. In contrast, the ether was a presuppositional posit: the success of the wave theory did not depend on waves propagating through a medium, and this notion was abandoned in the subsequent theory. Realists are, therefore, justified in believing that light exists as an electromagnetic wave but need not believe that ether exists as well. Another example comes from the caloric theory or phlogiston theory of combustion, where phlogiston is a presuppositional posit. It is correct that something is exchanged in combustion, even if the earlier theories were mistaken in thinking it was released rather than absorbed. This line of thinking can also be connected to Lakatos' view of research programs (1970), according to which theories have a hard core (the working posits) and a protective belt (which includes the presuppositional posits) that may be adjusted.

On a related note, we can mention *entity realism* (Hacking 1982), which promotes a realist interpretation of objects that can be manipulated. One is justified in believing in the existence of objects that one can experiment on—if you can spray them, they are real. A more sophisticated version of this view is *semi-realism* (Chakravartty 1998), which focuses on properties rather than objects. This approach distinguishes between causal properties—those that one can be a realist about—and "idle" properties, which are putative properties posited by the theory. Causal properties are in a causal connection with experimental apparatuses and can thus be detected, while idle properties are not directly connected to experimental observation. Additionally, there are other varieties of restricted realism, such as *eclectic* or *progressive realism*, proposed by Saatsi (2008), which focus on what in a theory is responsible for its empirical success (see also Chapter 3).

In contrast, other restrictions of realism that do not focus on objects or properties, but instead refer to the vaguer notion of structure, can generally be categorized as *structuralism*. Structuralism can be understood in different ways, with the key distinction being between *epistemic structuralism* and *ontic structuralism*. The epistemic version of structuralism suggests that realism about objects is unjustified, but we can still know the structure between them. In other words, the pessimistic meta-induction argument is correct in asserting that theory changes often involve discontinuity at the level of unobservable entities. However, most of the mathematical content of the old theory, which describes the structural relations between physical bodies, carries over to the new one. Therefore, realists may not be justified in believing what the theory says about the nature of physical objects, but they are justified in believing that the structure between these objects, which is preserved during theory change, is true. Thus, the pessimistic meta-induction challenge is solved because structure is preserved across theory changes. An example can be found in the wave theory of light and electromagnetism: the two theories disagree about what oscillates (the ether or the electromagnetic fields), but they agree that a wave equation describes the phenomenon in both theories. One could limit structuralism to epistemology, suggesting that we can only know about structure, not about objects. However, other types of structuralism are ontic, arguing that only structure fundamentally exists. These views are often motivated by the problematic nature of objects in quantum theory. There are several ways to express ontic structuralism. One radical view, called *eliminativism*, asserts that there are no objects—only relational structures. The objects, whether microscopic or macroscopic, are not fundamental but emerge from these relational structures (French 1998, Ladyman 1998). Another view, called *moderate structuralism*, holds that both objects and relations among them fundamentally exist, in contrast to eliminativism. According to this view, objects do not exist independently of the structure to which they are related. Rather, both objects and structure are equally fundamental, and objects are characterized solely by the relations in which they stand (Esfeld 2004). I will discuss structuralism within quantum theory in Chapter 2.

While structuralism has gained popularity in discussions of quantum mechanics, explanationism and other object-oriented types of realism have not had the same focus (with the possible exception of Saatsi's approach). *Quantum mechanics has long been considered incompatible with scientific realism.* While quantum theory is incredibly powerful in making precise predictions, it is also extremely difficult to interpret from a realist perspective. As discussed in Chapter 2, many have argued that quantum mechanics suggests that physical objects have contradictory properties, such as being in one place and not in that place at the same time, or that properties do not exist independently of observation. For a long time, this led many to conclude that scientific realism was impossible. Fortunately, the situation has changed. Today, various proposals for quantum theories allow for a realist interpretation, and *several realist frameworks are available.* Nonetheless, most examples from quantum mechanics are used to motivate ontic structuralism rather than other approaches. In quantum mechanics, for instance, the Schrödinger equation describes the temporal evolution of the wave function. The wave function is involved in deriving most, if not all, of the predictions and explanations that quantum theory provides. Within the framework of explanationism (or semi-realism), it seems that one should regard the wave function as a working posit (or a causal property) of quantum theory. However, these views would not be able to defeat the pessimistic meta-induction argument, since the wave function arguably does not have any classical analog, meaning there is a radical discontinuity between classical and quantum theory. This presents a challenge. In Chapter 5, however, I will show how one can extend explanationism to quantum theories when paired with a particular view about the metaphysics of quantum mechanics, as presented in Chapter 4. I will then compare this view to structuralism in Chapter 6 and argue that explanationism is preferable in several respects.

Super-Empirical Virtues

Against the challenge of underdetermination, realists typically argue that one is justified in selecting a theory over alternatives by appealing to something beyond mere empirical data. When discussing *empirical virtues*, people generally refer to attributes like empirical adequacy and empirical strength. A theory's *empirical adequacy* means that it makes correct predictions, while its *empirical strength* captures the theory's broad scope and its ability to make a large number of predictions. One can further distinguish between predictive power in breadth and in depth. *Predictive power in breadth* refers to a theory's ability to make predictions across various domains. For example, Newtonian mechanics has this type of power because it applies to both terrestrial bodies and planetary motions. In contrast, *predictive power in depth* means that the theory makes highly precise and accurate predictions within a given domain.

However, the realist maintains that these are not the only positive features a theory may possess. There are additional virtues, beyond just correct predictions, that may make a theory more worthy of belief. These characteristics are often referred to as *super-empirical* or *extra-empirical virtues*. Many realists believe that possessing these virtues justifies belief in a theory. For instance, among empirically equivalent theories, one is justified in preferring a theory that, all else being equal, is compatible with other well-established theories from different domains. This quality is often called *coherence*, a super-empirical virtue. For example, Kepler's laws of planetary motion are compatible with Newtonian mechanics, as the former can be derived from the latter (see the next section). In contrast, quantum theory and relativity have been argued to be in tension. This is partly due to the violation of Bell's inequality, which suggests that quantum theory is nonlocal—meaning, roughly, that influence can propagate instantly between arbitrarily distant events. As discussed in the end of the first section of this chapter, this nonlocality contradicts our intuitions. Moreover, this nonlocal behavior does not sit well with relativity theory, which asserts that nothing can travel faster than light. If this is indeed the case, as Einstein himself believed, then one might argue that a local alternative to quantum theory should be preferred, assuming it can be shown to exist. This would allow the two remarkable theories to be consistent with one another (see Chapter 2).

Another important super-empirical virtue is simplicity, which can be understood in various ways. One way is through the lens of *parsimony*, often associated with Ockham's razor: a theory is simpler if it postulates fewer entities or has fewer equations. For instance, Newton chose to explain light in terms of particles because this approach did not require the addition of any new entities to the world. Alternatively, simplicity can be considered in terms of *mathematical description*: a theory is simpler if its equations are less mathematically complex. Using this perspective, one could argue that Copernican astronomy is simpler than Ptolemaic astronomy, at least in its initial formulation, because it required fewer epicycles. In the Ptolemaic system, planets were thought to move in small circles (epicycles) that themselves moved along larger circles (deferents). Others have argued that the many-worlds interpretation of quantum theory is simpler than its alternatives because it involves fewer and mathematically simpler equations (see Chapter 3). The tradition of viewing simplicity as a theoretical virtue is long-standing, with notable proponents including Aristotle, Kant, Galileo, and Newton. For instance, Newton used simplicity to argue that the same force acting on Earth must also act on other planets. Many scientists, most notably Lavoisier and Einstein, believed that the aim of science is to cover the greatest number of phenomena with the smallest number of hypotheses (see Wallace 2012, French 2014).

Another extra-empirical virtue is *fruitfulness*: a theory can be considered fruitful if it leads to new discoveries or the development of new theories. Closely related to fruitfulness is *plasticity*, which refers to a theory's ability to adapt its formalism to various contexts. For example, some have argued that the Newtonian formalism is less suited than the Lagrangian or Hamiltonian formalisms for extension to other theories. Relativistic quantum field theories, for example, are typically formulated starting from a Lagrangian or Hamiltonian. Because of this, it has been suggested that theories that naturally take on a Lagrangian or Hamiltonian formulation are more amenable to relativistic extensions, making these formulations preferable to Newtonian ones. Additionally, one might consider certain quantum theories, like spontaneous localization theories, to be more fruitful than others, such as pilot-wave theory, because they appear less in conflict with relativity theory (though, see Chapter 4).

Another virtue worth mentioning is *naturalness*. One way of understanding this is in terms of *lack of fine-tuning*, meaning the ingredients of the theory should not be cherry-picked or excessively tailored to fit a specific set of observations. Ideally, a theory should explain phenomena in a general way, not being overly sensitive to specific details. This is connected to the notion of *typicality*, as follows. Assume one has a microscopic theory like Newtonian mechanics. The ideal situation would be to explain macroscopic phenomena in terms of microscopic Newtonian dynamics for all possible initial conditions. This can sometimes be done, as for example with rigid bodies, as outlined in the next section. However, in the case of gases, this is not strictly possible. One can show that phenomena are reproduced only for most initial conditions, and if so, these phenomena are often referred to as *typical*. This is still considered a good result because when a phenomenon is typical, we expect it to occur, and we are not surprised when it does. This contrasts with phenomena that occur only under very special or highly specific initial conditions. We can account for such phenomena by reconstructing where they came from, but can we truly say we've *explained* them? (For more on typicality and explanation, see "Further Readings.")

Be that as it may, in general, it is *not obvious how to define these virtues univocally* (see also Chapter 4). Moreover, they may conflict with one another: a simple theory might be less fruitful than another. For instance, one can arguably maintain that the Newtonian formulation of classical mechanics is simpler than the alternatives, even if it might be thought of as less fruitful. Similar considerations can be drawn in the comparison between quantum theories: I argue in Chapter 4 that the pilot-wave theory, even if arguably less fruitful, is simpler and more explanatory than spontaneous localization theories. In addition, even if one could uncontroversially define and unquestionably rank these features in order of importance, it would still be unclear why possessing these virtues would make the theory more likely to be true. *Why should we believe, say, that simplicity is evidence of truth*? Realists have provided multiple responses to this challenge, which I am not going to evaluate here (for references, see "Further Readings"). According to some, if there is no reason to favor simplicity, there is no reason to favor nonsimplicity also. In other words, while it may not be the case that a simple theory guarantees the theory's truth, *ceteris paribus*, it seems more convenient to use simple theories than complicated ones. In this sense, simplicity is a pragmatic virtue, but a virtue, nonetheless. Alternatively, some claim that *a priori*, nature is simple, or that it is self-evident that simplicity is a guide to truth. Others maintain that simplicity has intrinsic value as a theoretical goal, for either primitive or aesthetic reasons. Some others contend that simplicity can be derived from some *rationality principle*, such as unification: a simpler theory is one which has fewer explanatory patterns. Or it can be justified inductively: since simplicity in the past has led us to approximately true theories, it is likely to do the same in the future. Otherwise, one can think of it as an explanatory virtue, or think that a simpler theory has less chance of being wrong. Regardless, thinking of simplicity as a theoretical virtue is common practice, as for instance by the proponents of the Humean's best system account (see Chapter 2).

Regardless, the virtue I consider most important for realism is *explanatory power*, because for realists, explanation is the essence of science. One should always prefer a theory with greater explanatory power, even if another theory is ontologically simpler or more elegant but less explanatory. For instance, classical mechanics, understood in Newtonian terms, postulates forces between particles with different properties, while in Lagrangian terms, there are no forces. So, while the Lagrangian formulation may be simpler, it is arguably less explanatory of the phenomena we observe, given that it is expressed in special coordinates rather than the "usual" three-dimensional ones. Still, explanatory power is (unsurprisingly) controversial to define, as different people have different ideas about what counts as a good explanation. In the next section, I discuss two types of explanations using the distinction between constructive and principle theories, arguing that constructive explanation is better, as it explains why something happens rather than merely specifying what we should expect to happen.

The Classical Constructive Schema for Particles

The distinction between theories of *principle and constructive theories* was originally proposed by Einstein (1919), who argued that science should aim at constructive theories, while principle theories are only provisional (for details, see Giovannelli and Howard 2025). Arguably, in his view, a constructive theory is one in which macroscopic objects (broadly understood) are constructed from microscopic (three-dimensional) entities, and in which the properties of what we observe and their behavior are accounted for dynamically. This means specifying how the microscopic fundamental entities evolve in time and interact with one another. In a constructive explanation, as defined here, there are two components: an ontology for matter and its law of evolution. The macroscopic objects of our everyday experience are considered to be composed of the microscopic entities in the ontology, and the macroscopic properties are dynamically reduced (under suitable approximations) in terms of the microscopic dynamics. Thus, *a constructive explanation assumes both compositionality and dynamical derivability*. Classical mechanics, as discussed earlier, follows this approach. I call the constructive techniques of explanation used in classical mechanics the "classical explanatory schema." Earlier, we saw how Newton proposed that light was made of particles and suggested an explanation of the laws of geometrical optics based on the motion of particles. However, this explanation was eventually disproven by experimental results (e.g., the observation of interference and diffraction of light), which indicated that the particle-based ontology was inadequate. This led to the acceptance of the wave theory of light. Newton's ontology and laws aimed to account for all phenomena and phenomenological theories: the orbits of planets, the trajectories of projectiles, the laws governing gases, the solidity of matter, the transparency of water, and even the color of the rainbow. This constructive, bottom-up picture assumes that the behavior of everything can be explained once the dynamics of the particles (the fundamental building blocks of matter) are specified. In simpler terms, constructive theories assume a Lego-bricks-style reductionism (properly understood). All macroscopic bodies (solids, liquids, and gases) are considered to be composed of microscopic particles interacting with one another in different ways. The laws governing their motions can be recovered (under suitable approximations) from the motion of the fundamental particles, for most initial conditions. This is the approach taken in statistical mechanics, and in some cases, like rigid bodies, fluids, Kepler's laws of planetary motion, and Galileo's law of free fall, it works even for all conditions. Here are some concrete examples of (arguably) successful explanations within the Newtonian framework. For solids, which do not change in shape or volume, rigid body dynamics was developed to extend the Newtonian framework. Fictionally, rigid bodies can be thought of as if their mass is concentrated at a single point—the center of mass. Simple equations for their motion can then be derived. These equations are theorems: they are derived from the fundamental Newtonian laws under appropriate approximations and, as such, express nonfundamental laws.

One striking success of Newton's theory using these techniques is its reproduction of Kepler's laws of planetary motion. Under the approximation that the masses of the planets are negligible compared to the mass of the Sun, the equations of the more fundamental theory (classical mechanics) reproduce Kepler's laws. The same is true for Galilei's law, which states that all bodies in free fall in Earth's gravitational field have the same acceleration, provided the height of the fall is negligible compared to the radius of the Earth. Similarly, fluids are extended composite bodies that can change shape depending on the container they are in, but not in volume (they are incompressible). For fluids, the situation is more complicated, but arguably, one can still derive a set of empirically adequate nonfundamental laws from the microscopic Newtonian motion. This is known as hydrodynamics. The case of gases is trickier, as they can change both in shape and in volume. Thermodynamics phenomenologically describes laws for gases, outlining regularities in their behavior. For example, if you increase a gas's pressure while keeping its volume constant, the temperature will rise. Similarly, the second law of thermodynamics states that a given quantity called entropy never decreases in isolated systems. A constructive explanation of these laws would involve deriving them from the classical microscopic laws governing the particles composing the gas. However, this is particularly difficult in the case of the second law, which shows a preferred direction: entropy never decreases over time. This temporal directionality is in tension with Newtonian mechanics, which does not display such a preference (see toward the end of this chapter for more). Nonetheless, Boltzmann demonstrated that all the laws of thermodynamics could be derived from the microscopic dynamics. Specifically, he showed that the second law should be understood statistically: entropy does not always increase, but it most likely does, or remains the same. More precisely, one could say that entropy does not decrease from most initial conditions. It is outside the scope of this book to explore the foundations of statistical mechanics and the meaning of probability in this context (see "Further Readings" section). I mention this here only to highlight that this was the constructive project initiated by Newton, motivated by his realist perspective. According to Newton, matter is made of particles—solids, liquids, and gases. The nonfundamental laws governing these composites, including all the properties they possess, can, in principle, be derived from the fundamental laws of classical mechanics.

In contrast, principle theories constrain physically possible phenomena using general principles. In other words, while in constructive theories everything is supposed to follow from the fundamental entities and their laws, *principle theories specify what can or cannot happen in terms of overarching constraints.* The type of explanation offered by principle theories differs from that of constructive theories. It is not framed in terms of microscopic particles moving and interacting, nor are macroscopic laws reduced to microscopic dynamics. (This is why principle theories are sometimes called *kinematic* theories, while constructive theories are referred to as *dynamical* theories.) Instead, principle theories explain why something happens—or does not happen—by appealing to whether it is permitted or prohibited by one of the principles. Examples include geometrical optics and thermodynamics. For instance, thermodynamics posits several key principles, such as "energy is conserved" and "entropy never decreases." These principles rule out certain phenomena: a process that violates energy conservation or causes entropy to decrease is physically impossible and thus cannot be observed.

Einstein, it is often argued, believed that *constructive theories provide a superior form of explanation and viewed principle theories as often provisional*—offered in the absence of a constructive alternative. Others, as we will see in Chapter 3 when discussing the information-theoretic approach to quantum mechanics, have challenged this view and argued that principle theories should instead be preferred. My own reason for favoring constructive explanations is that they account for macroscopic phenomena by specifying the mechanisms that give rise to them in terms of a fundamental material ontology and its laws. In contrast, principle explanations offer expectations about what is possible, i.e., what will happen is simply what is allowed by the principles. However, *they do not explain why the phenomenon happens*—there is no underlying mechanism to account for our observations or for why the principles themselves hold. To be clear, I do not believe that proponents of constructive explanations dismiss principle explanations outright. Rather, principle theories are considered provisional—they *cry out* for a deeper account. They still owe us an answer to the question: *Why do the principles hold?* Moreover, there is a sense in which constructive explanation is the simplest: it makes the fewest assumptions, is easy to grasp, and is visualizable—Lego-like in nature. It also remains consistent with the type of explanation offered by our best historical scientific theories. So, all else being equal, there seems to be no good reason to abandon it. In any case, in Chapter 3 I will also show that accepting principle theories comes with a significant consequence: it commits one to a much weaker form of realism—one that many (if not most) realists would likely find unsatisfying. Therefore, *ceteris paribus*, I argue that constructive explanations should be preferred for their greater explanatory power.

The Classical Constructive Schema for Waves

As we have just seen, the classical explanatory schema—when applied to a microscopic, localized ontology such as particles—offers an explanation for why the regularities described by principle theories like thermodynamics hold. That is, macroscopic objects are not only composed of microscopic fundamental entities, but the macroscopic laws themselves can be derived (under suitable approximations and on appropriate scales) from the microscopic dynamics. Since this will become relevant in later discussions, let us now consider what it would mean to have *a constructive explanation* based on a different kind of fundamental ontology: *waves* rather than particles. In classical mechanics, as proposed by Newton, the ontology is purely particle-based. But what would be required of a theory that postulates only wave-like fundamental entities? Such a theory would need to explain why matter—now conceived as fundamentally wave-like—nevertheless appears to behave as though it consists of small, localized "crumbs" that leave distinct tracks in detectors. This idea of a wholly wave-based ontology was first proposed in the quantum domain by Schrödinger (see Chapter 2), who suggested that what we take to be particles are actually well-localized *wave packets*, formed through the superposition of waves with different wavelengths. We misidentify these as indivisible, point-like, structureless particles, but if examined closely, their composite wave-like nature would become apparent. In this way, even a wave ontology can accommodate a kind of compositionality—provided the waves oscillate in the same three-dimensional space where we perceive particles. According to the superposition principle, waves can combine to form localized wave packets. However, because particles are observed to follow stable trajectories, such wave packets would need to remain localized over long periods and not disperse or interfere with themselves. In other words, to serve as a constructive explanation, one would have to show that—at least for all practical purposes and in the appropriate limits—the fundamental wave equation (in the case we will examine next, the Schrödinger equation) has solutions that approximate the classical Newtonian equations of motion.

As we will see in Chapter 2, this was precisely what Schrödinger tried to achieve, but ultimately failed to do. More generally, this is the goal that any constructive theory with a wave-based fundamental ontology should strive toward. To anticipate the discussion in Chapter 3, this is also what the view called *wave function realism* would need to accomplish in order to provide a proper constructive explanation—but, as I will argue, it fails to do so.

Now consider the case of a *dual fundamental ontology*: particles and waves. The rise of the wave theory of light, along with the development of electromagnetism, led many to believe that there are two fundamental kinds of entities in the world—matter, composed of particles, and light, which consists of waves. In this framework, the waves represent radiation (light), not matter, so their behavior does not need to reproduce classical particle trajectories in any specific limit. Electromagnetic waves are delocalized entities, and compositionality, strictly speaking, does not apply to them. Nevertheless, there is a sense in which the explanations provided by electrodynamics follow *the same constructive logic* as those in Newtonian mechanics. Just as Kepler's laws of planetary motion emerge as an appropriate limit of Newton's equations, the laws of geometrical optics arise from Maxwell's equations under the eikonal approximation. To classify a theory as constructively explanatory, one must be able to recover the laws and properties of macroscopic objects under some suitable approximation from the fundamental dynamics. Beginning in Chapter 4, I argue that in the case of classical electrodynamics, there is, in fact, no genuine dual ontology. Only one of the two—particles—should be taken as fundamentally representing matter. The other—fields—is more appropriately understood as describing interactions among particles. Similar considerations apply to quantum theory, which should also be viewed as a theory of localized (or effectively localized) entities. In this way, compositionality remains intact, and the macroscopic laws can still be dynamically derived from the microscopic dynamics.

As a final remark, it is worth noting that *some have associated causal, mechanistic explanation with constructive theories, and unificatory or structural explanations with principle theories* (Flores 1999). These are intriguing connections, but I will not explore them further here. I would simply note that, contrary to this common association, there is a parallel between constructive theories and Hempel's deductive-nomological model (Hempel 1952), which conceives of explanations as arguments with law-like premises. Since Kitcher's unification model (1981) can be viewed as an extension of the deductive-nomological model, it is plausible that constructive explanations—at least as I have described them—are compatible with the unification model as well.

Symmetries and Perspective Free Descriptions

There is a sense in which the explanatory power of a theory is connected to its simplicity—an idea we have not yet explored. This is the focus of the current section. One might think that a theory is simpler if it exhibits symmetries, meaning that its laws describe phenomena as the same from different points of view. In what follows,

I will discuss the *importance of symmetries* and their relation to other theoretical assumptions, such as the uniformity of nature.

Many postulates and assumptions are (to varying degrees) ingrained in the development of scientific theories. One prominent example is the assumption that *nature is uniform*. The rationale for this assumption is that, without it, it would be difficult to conceive of universal laws of nature—laws that are timeless and valid everywhere. If we assume that nature is indeed uniform, then our theories ought to reflect this feature. To understand what this entails, we must examine the notion of symmetry. The most general definition of symmetry is *a transformation of an object that leaves some aspect of it unchanged*. The unchanging aspect is said to be *invariant* under the symmetry. The object in question can vary widely—it may be a geometrical figure, all of space (or spacetime), or even an equation. Symmetry transformations can be continuous, if they resemble a smooth or gradual change, or discrete, if they involve a more abrupt change, like a flip. For example, consider a geometric figure like a sphere. A sphere is symmetric under rotation: turning it around any axis leaves it indistinguishable from its original orientation—one cannot tell whether they are looking at the original or the rotated version. A useful distinction often made is between passive and active transformations. These differ only in perspective: in a passive transformation, the observer's frame of reference changes while the object remains fixed; in an active transformation, the object itself changes while the frame of reference stays the same. Now consider symmetries of space and time. A continuous symmetry of space is translation: space remains the same regardless of whether the "center" is here or there. The same applies to time translations—time has no intrinsic "center" or "beginning." The invariance of physical laws under such space and time translations is known as the *homogeneity* of space and time, while rotational symmetry of space is referred to as spatial *isotropy*.

These symmetry properties of space and time can be understood in two ways: as revealing fundamental features of spacetime itself (as suggested earlier), or as expressing the idea that *physical phenomena do not depend on where they occur (spatial translation symmetry), when they occur (temporal translation symmetry), or in which direction we observe them (spatial rotation symmetry)*. In this latter sense, these symmetries are not just features of spacetime, but of the theories used to describe the phenomena. Specifically, they become symmetries of the fundamental equations governing physical evolution. A theory is said to be symmetric or invariant under a transformation if applying that transformation *leaves the fundamental equation unchanged* in a way that preserves the same physical behavior. (Sometimes, this is phrased by saying that the equation is *covariant* under the transformation: it changes in form, but still describes the same physical reality.) For instance, Newtonian mechanics is invariant under spatial translation: if we replace r with $r+c$, where c is a constant vector, in the equation $m\frac{d^2r}{dt^2}=F$ (where the force depends only on the relative distances between objects), the form of the equation remains the same. This implies that the phenomena it describes are unaffected by the shift in position. The same invariance holds for time translation, spatial rotation, and even time reversal: if we reverse time by replacing t with $-t$, the equations of motion still describe a physically possible behavior (see Figure 1.2). This means that under ideal conditions, physical processes are time-reversal symmetric. For example, if we film the motion of billiard balls on a frictionless table and play the footage in reverse, the backward motion still appears physically plausible.

Symmetry properties are almost universally regarded as *essential requirements* for a physical theory. Scientists often use symmetry constraints to construct new theories, and they tend to prefer invariant (or more invariant) theories over noninvariant (or less invariant) ones. One might think this preference stems from the fact that symmetric theories are more aesthetically pleasing or mathematically convenient. However, these are *pragmatic considerations* and thus of limited interest to the realist, for it is unclear why nature should favor more beautiful or mathematically elegant theories. Another reason to value symmetries lies in their *connection to conserved quantities*, as established by Noether's theorem. According to this theorem, every continuous symmetry corresponds to a conserved quantity—for instance, time translation symmetry leads to energy conservation, and spatial translation symmetry leads to the conservation of linear momentum. The thought is that if we value conserved quantities, we should also value symmetries. But again, it is not obvious why conserved quantities should be valued in the first place, except perhaps for similar pragmatic reasons (simplicity, elegance, etc.).

In my view, the most compelling reason to require symmetry in a theory stems from the idea that *a theory capable of describing a phenomenon as neutrally as possible offers stronger evidence of its truth.* That is, if a theory represents phenomena in a way that is maximally independent of our particular perspective, then it seems more likely to be correct. Symmetry, in this sense, is a *desideratum* for the realist because it decouples

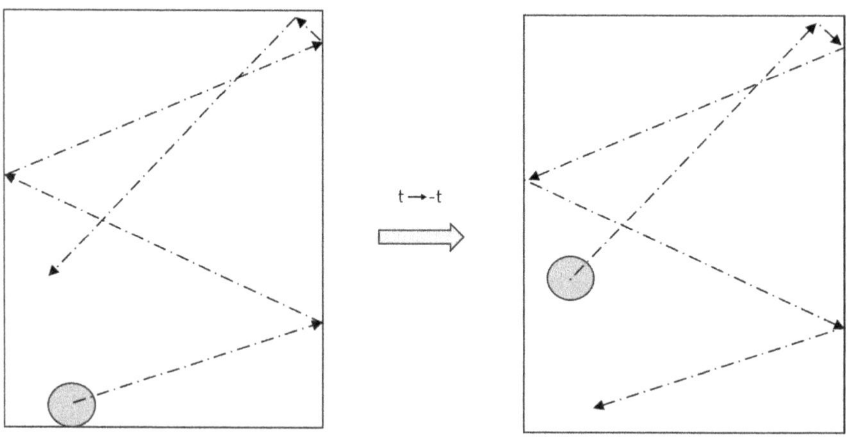

Figure 1.2 Time reversal in classical mechanics.

the phenomena from our description of them. It supports the idea that the theory is getting at something objective. Let me elaborate. When we describe a phenomenon, we often use reference frames because doing so is convenient—it helps relate one phenomenon to others or to specific locations. For example, it may be convenient to say that a hurricane is approaching a point 5 kilometers west of Chicago, Illinois, so people nearby can assess whether they're at risk. But that same hurricane would remain exactly the same if I described it as approaching a point 7,297 kilometers west of Milan, Italy. Likewise, it makes no difference whether we call it "hurricane" in English or "uragano" in Italian—the phenomenon itself does not change. In other words, the hurricane is what it is, regardless of how we describe it. It is *perspective-independent.* By the same logic, the laws of a physical theory should ideally be independent of the specific perspective from which the phenomena are observed. That is why symmetry should be a requirement for any realist theory: it ensures that the theory's description of reality does not depend on arbitrary features of the observer's frame of reference.

This idea has a *long-standing tradition.* In classical mechanics, there is a physical distinction between accelerated frames and nonaccelerated (or inertial) ones. However, all inertial frames are indistinguishable from one another. For example, an inertial frame could be a rest frame—such as when I observe a train moving at a constant velocity v while sitting on a bench at the station. But another inertial frame could be me observing the train from my bicycle, which is also moving at a constant velocity relative to the station. Similarly, I remain in an inertial frame if I'm inside the train and walking uniformly along the cabin. In each of these cases, the motion of the train can be described using the same equation: $F = ma$, where acceleration is zero. This indistinguishability of inertial frames is captured by the fact that Newtonian mechanics is invariant under transformations corresponding to uniform motion—so-called Galilean boosts. When one substitutes r with $r - vt$ where v is constant, the equation $F = ma$ remains unchanged. Since acceleration is the time derivative of velocity, and v is constant, the acceleration remains unaffected. This is the content of *Galilean relativity* in classical mechanics: the laws of motion are the same in all inertial (i.e., unaccelerated) frames. However, this is not true for accelerated frames. If I am accelerating relative to the train—or vice versa, such as when the train takes a turn (resulting in radial acceleration)—then the situation changes. The equation $F = ma$ no longer accurately describes what is observed unless one introduces *apparent* (or fictitious) forces. These forces do not arise from physical interactions, such as gravitational forces do (which depend on mass), but are instead added to Newton's law in order to make it applicable in accelerating frames. This illustrates that while Newtonian mechanics is invariant under Galilean transformations—meaning its laws hold in all inertial frames—it is not invariant under arbitrary transformations involving acceleration. The laws of mechanics are not the same in *all frames*, only in inertial ones.

So, it seems to me that the reason *we care about symmetries is because we value perspective-free descriptions.* To support this, consider Einstein's formulation of *special relativity.* He proposed two principles: the constancy of the speed of light and the relativity principle. The latter is the same principle found in Galilean relativity in classical mechanics—that all laws of nature should be the same in all inertial frames. However, combining this principle with the constancy of the speed of light led to a theory that is no longer invariant under Galilean transformations (i.e., $r \to r - vt, t \to t$). To see this, consider again Galilean relativity: the phenomena appear the same to us sitting on a bench at a train station as they do to someone on a train moving at a constant velocity v relative to us. Furthermore, if a person on the train gets up and walks at a constant velocity u, then from the station's perspective, we would observe them moving at $u + v$. In classical mechanics, for the laws to remain valid across frames, velocities must add vectorially. This, however, is not the case in special relativity. According to Galilean transformations, a beam of light from a flashlight held by a passenger on the train would appear to us as moving at speed $c + v$, greater than the speed of light c. But this contradicts the postulate of the constancy of the speed of light: c must remain constant in all inertial frames. To reconcile this with the relativity principle, Einstein adopted a different set of transformations—*Lorentz transformations*, which had already been discovered by Lorentz. These transformations replace Galilean ones and preserve the invariance of physical laws across inertial frames, including the constancy of the speed of light. They involve a more complex relationship between the coordinates of two systems, including changes in temporal coordinates. $x \to \frac{(x - vt)}{\sqrt{1 - \frac{v^2}{c^2}}}$, $t \to \frac{\left(t - \frac{vx}{c^2}\right)}{\sqrt{1 - \frac{v^2}{c^2}}}$. In other words, to maintain the invariance of physical laws across inertial frames, Einstein required Lorentz, rather than Galilean, invariance. In fact, it has been suggested that relativity theory should be called the *theory of invariants*, since its core idea is that physical laws must remain invariant under transformations between frames. Einstein clearly prioritized this principle above all: the guiding idea behind his theory of relativity is that *the laws of physics are the same for all frames*, inertial or not. That is, there is no privileged point in spacetime from which to derive a "true" description—there is no privileged vantage point.

The view I defend in Chapter 4 provides a constructive explanation based on an ontology that preserves symmetries. By supplying the most perspective-free constructive explanation possible, my approach offers the best explanation of the phenomena.

Further Readings and Discussion

Further reading on metametaphysics—that is, the nature of metaphysics and its relationship to science and physics—can be found in a wide range of articles and books. For instance, Wasserman et al. (2009) and Bliss and Miller (2021) offer valuable collections on metametaphysics, while Tahko (2015) provides an informative, albeit brief, introduction. Different perspectives on the relationship between metaphysics and science are explored in Lowe (1998), Ladyman and Ross (2007), and Sider (2011,

2020). Several books in the philosophy of physics directly engage with metaphysical questions. These include Albert (1996), Maudlin (2007a), Lewis (2016), Ney (2021), and North (2021a). An invaluable and continually updated resource is the *Stanford Encyclopedia of Philosophy (SEP)*: https://plato.stanford.edu/. The entry on metaphysics by van Inwagen and Sullivan (2021) is especially noteworthy. A clear and concise introduction to metaphysics is also found in Ney (2014).

For those interested in scientific realism and antirealism, the SEP provides excellent bibliographies. The entry on scientific realism (Chakravartty 2017) offers a comprehensive overview of realist perspectives. Psillos (2022) discusses the pessimistic meta-induction and various responses to it. Stanford (2021) provides a general review of the surrounding debates, while Niiniluoto (2019) explores questions of scientific progress, its relation to rationality, explanation, and truth. For discussions of simplicity as a super-empirical virtue, see Baker (2022). A treatment of constructive empiricism is provided in Monton and Mohler (2021). Notable recent work on realism includes Psillos (1999) and Chakravartty (2007), while important contributions to antirealism come from Stanford (e.g., 2006), Rowbottom (2019), and, of course, van Fraassen (1980). Restricted realist strategies—such as structural realism and explanationism—will be discussed in Chapters 2 and 4, respectively, where further bibliographical information will also be provided.

I also touched on debates within the philosophy of physics, particularly regarding the metaphysics of classical mechanics, classical electrodynamics, and statistical mechanics. On classical mechanics, see Wilson (2013), Curiel (2014), Barrett (2015), and North (2021a, 2021b). Discussions of the ontology of classical electromagnetism can be found in Frisch (2008, 2009), Allori (2015a), Lazarovici (2018), and Hubert and Hartenstein (2021). (See Chapter 4 for more on the status of electromagnetic fields.) Important texts on the foundations of statistical mechanics include Sklar (1993), Albert (2000), and Bricmont (2022). For work on typicality and its connection to explanation, see Goldstein (2001), Volchan (2007), Lazarovici and Reichert (2015), Wilhelm (2019), Allori (2020a, 2022a), Badino (2020), Crane and Wilhelm (2020), Lazarovici (2020), Maudlin (2020), and Hubert (2021). For critiques of this approach, see especially Frigg (2009, 2011) and Frigg and Werndl (2012).

The distinction between principle and constructive theories has been taken up in Flores (1999), Balashov and Janssen (2003), Brown (2005), Brown and Pooley (2006), Felline (2011), Van Camp (2011), and Giovanneli and Howard (2025). For a recent and comprehensive defense of principle-based explanation, see Silberstein, Stuckey, and McDevitt (2018).

Finally, the SEP entry on symmetries and symmetry breaking (Brading et al. 2021) provides a detailed overview of various notions of symmetry and their role in scientific practice. Other important discussions can be found in Brading and Castellani (2003), Baker (2010), and Dasgupta (2016).

2

Quantum Mechanics and the Structuralist Revolution

Since its introduction, quantum theory has presented *a mystery from a realist perspective*. It is highly successful in making predictions, but it is difficult to understand what it tells us about the nature of the world. Often, it is claimed that quantum theory implies particles behaving like waves, waves behaving like particles, observers creating reality, logic failing, and so on. While the specifics may vary, quantum theory is frequently depicted, in Kuhnian terms, as *a scientific revolution*—similar to, or even more radical than, the shift from Ptolemaic to Copernican astronomy. It is argued that the worldview provided by quantum theory is vastly different from that of classical mechanics. Some claim this revolution brought a paradigm shift, suggesting that we must completely change how we view the world and science itself. Furthermore, it is often asserted that *this change is necessary*, with no alternative, less radical understanding of quantum phenomena. Initially, some even argued that quantum theory made scientific realism impossible. I argue, however, that a much less radical understanding of quantum mechanics is possible and, in fact, preferable—a position I begin to develop in Chapter 4. In this chapter, I discuss some of the *challenges to realism* that have arisen from quantum theory. The responses to these challenges have led to the development of *structuralism*, which is also addressed here. However, this should not be seen as a comprehensive overview of the structuralist perspective. I focus primarily on the aspects relevant to comparing it with my approach. Additionally, I mention other approaches to quantum ontology that share notable features with structuralism, including the information-theoretic approach, Qbism, and pragmatist approaches.

Chapter Schema Details

The first section briefly reviews the main components of quantum theory as presented in physics textbooks. I refer to this theory as "*standard quantum mechanics*" to distinguish it from other quantum theories that were later proposed, which I discuss in the third section of this chapter. It is important to note that the presentation here is not a comprehensive mathematical treatment of the theory; rather, I focus on the material most relevant to questions concerning realism. In the second section, I address some challenges to realism within the quantum domain, particularly those related to concerns about individuality arising from distinctively quantum features such as entanglement. In other words, some aspects of standard quantum theory have led people to suggest that the concept of an object with definite properties must be rethought or even abandoned. As we will see, this challenge naturally gave rise to *ontic structuralism* (which I will refer to simply as "structuralism") as a way to preserve realism in the face of these issues. Finally, I examine *several problems* with structuralism and the type of explanation it seems to imply. This allows me to argue that, while many objections to structuralism have received suitable responses, the account remains unnecessarily complicated in an important way. From a realist perspective, I believe the most serious weakness of this viewpoint is its explicit *renunciation of constructive explanation*. In the fundamental ontology, we are left with only structure, while objects—both microscopic and macroscopic—emerge as useful patterns in a top-down, rather than bottom-up, approach. Structuralists explicitly acknowledge this aspect of their proposal and argue that this is precisely how it should be. However, I contend that this makes their version of *realism too weak*, bordering more on instrumentalism than on the type of realism a realist might desire, especially if alternatives are available.

Standard Quantum Mechanics

Classical mechanics and relativity theories were each developed by a single individual—Newton and Einstein, respectively—who had a clear metaphysical hypothesis in mind about what should represent matter in their theories when they proposed them from scratch. In contrast, quantum mechanics emerged from the work of many scientists and primarily arose as an attempt to reconcile some recalcitrant experimental evidence into a unifying framework, without a clear underlying ontology in mind. This difference is part of the reason why, unlike other theories, quantum theory is often introduced historically, by presenting how certain data could not be explained within the classical framework. A *historical introduction* is essential for understanding why we now have standard quantum mechanics. Indeed, in Chapter 7, I will argue that several contingent historical events, rather than rational arguments, have been crucial in shaping what we today accept as standard quantum mechanics. However, I will also argue that this framework is not the best possible theory for explaining quantum phenomena from a realist perspective. Thus, before delving into the formalism of the theory, let me briefly provide some historical background.

A Short History of the Standard Formalism

To cut a long story short, the classical framework discussed in the previous chapter, in which matter is made of particles and light is a wave, dominated unchallenged until the early twentieth century. The main problem for classical mechanics that led to the development of quantum theory came from the *study of atomic structure*. Several experiments involving the recently discovered charged particles, such as electrons and protons, along with the study of radioactivity in the late nineteenth and early twentieth centuries, led to the hypothesis that fundamental particles—originally thought to be indivisible (hence the name "atoms," derived from the Greek prefix *a* —meaning "not," and *émnō/τέμνω* meaning "to cut")—actually had an internal structure. The problem was to understand how various charged particles could be arranged in a single, neutrally charged atom. A natural proposal suggested that negatively charged electrons orbited a positively charged nucleus containing protons and neutrons. However, this model conflicted with electromagnetism, which predicted that such an atom would be unstable. Accelerated particles lose energy, causing their orbital radius to slowly decrease until they ultimately collapse into the nucleus. Bohr proposed an atomic model to solve this problem, introducing what would later be called a quantization rule: only certain orbits are physically accessible to the electron. In other words, the possible values for the electron's orbital radii are discrete, or *quantized*. This hypothesis solved the problem of the atom's stability and could also explain experimental data, such as the atomic spectra—the set of frequencies of light emitted or absorbed by different atoms. Moreover, new phenomena were discovered, leading to the assumption that matter has a new fundamental, genuinely quantum property, namely *spin*, and that other quantities, like energy, are also quantized. The initial challenge scientists faced was to find the corresponding set of quantization rules, much like Bohr's model. However, the resulting theory—sometimes referred to as "*old quantum mechanics*"—was essentially a supplementation of classical physics with quantization rules, but without any underlying theoretical understanding.

A natural approach might have been to develop a new theory, based on classical ontologies of particles and waves, to account for the quantization rules from a constructive perspective, as discussed in the previous chapter. In this view, the quantization rules would arise as a consequence of the dynamics and interactions between the particles in the atom and the waves constituting light. However, new experimental and theoretical discoveries suggested that such an understanding was not possible (see Chapter 7 for more). It was argued that all the classical ontologies used to explain quantum phenomena led to *contradictions*, requiring a *complete revision* of the scientific paradigm. On the one hand, observations of blackbody radiation frequencies (where a black body, like a star, emits light at different frequencies depending on its temperature) and phenomena such as the photoelectric effect (where electrons are emitted by a metallic surface exposed to visible light) and the Compton effect (the scattering of electromagnetic waves on matter) led to doubts about the established assumption that light was a wave. In fact, all the data from these experiments were incompatible with the predictions based on the wave theory of light. Instead, they were explained by postulating a particle ontology for light: in 1905, Einstein proposed that light consists of particles, which he called "photons." This hypothesis also successfully explained the other experimental results. On the other hand, some experiments showing diffraction and interference patterns produced by objects previously understood as particles, such as electrons, challenged the particle view of matter and supported the wave hypothesis.

As a result, given the mutual incompatibility of particles (which are localized) and waves (which are spread-out entities), most scientists decided to *set aside ontological questions* and began developing new theoretical frameworks to systematically reproduce the experimental data. A first attempt to make old quantum mechanics more cohesive and unified came in 1925 with *matrix mechanics*, proposed by Heisenberg, Born, and Jordan. They introduced systematic tables (matrices) of allowed values for common physical quantities, known as "observables," such as energy, position, or velocity. These matrices were interpreted in terms of specific *linear operators*—self-adjoint or Hermitian operators. (An operator, usually indicated by a "hat" on top of a capital letter, such as $\widehat{A}$, is a function that transforms vectors into other vectors. A linear operator preserves linear combinations: if a vector is a linear combination of other vectors, their transformed versions are combined in the same way. See below for self-adjointness.) The idea was to interpret these operators *as representing the quantized properties* of physical systems (discussed further below). Observable properties were associated with operators through a method called Dirac's first quantization, based on a formal analogy with Hamiltonian mechanics. This involved substituting Poisson brackets in Hamilton's equations with the commutators of the corresponding operators: $\{f,g\} \to \frac{1}{i\hbar}[\widehat{f},\widehat{g}]$, where $h = 6.62 \times 10^{-34}\, m^2\, kg/s$ is Planck's constant and $\hbar = h/2\pi$ is the reduced Planck's constant. (The Poisson brackets will not be defined here, but the commutator is given by $[\widehat{f},\widehat{g}] = \widehat{f}\widehat{g} - \widehat{g}\widehat{f}$. Classically, for position x and momentum p_x, one has $\{x, p_x\} = 1$, so the corresponding operators satisfy: $\left[\widehat{x},\widehat{p}_x\right] = i\hbar$. From this relation, it follows that the position operator is the multiplication operator, and the momentum operator is the differential operator. All other operators follow from these basic operators.)

In this way, to use the terminology introduced in the previous chapter, matrix mechanics is arguably best understood as a *principle theory*, in which the quantization rules and the operator formalism serve to systematize phenomena and exclude what is physically impossible. No hypothesis regarding an underlying ontology was initially proposed (though see below), so constructive explanation was not part of the original approach. The following year, Schrödinger—selectively building on de Broglie's earlier work—provided the first attempt at a *constructive quantum theory*, based on a wave ontology for matter, suitably dubbed *wave mechanics*. Schrödinger proposed an equation (which now bears his name) to describe the evolution of what he called the *wave function*. This equation could explain the quantization rules in oscillatory terms, namely as stationary nodes of the wave function, which he (provisionally) took to be the ontology—that is, the object meant to represent matter in wave mechanics. His project aimed to show that all quantum phenomena could be understood in terms of such a wave, following the classical explanatory framework discussed in the previous chapter. This included the idea that the common assumption—that matter is fundamentally made of particles—was mistaken. From a more fundamental perspective, as anticipated in Chapter 1, what we perceive as particles are instead

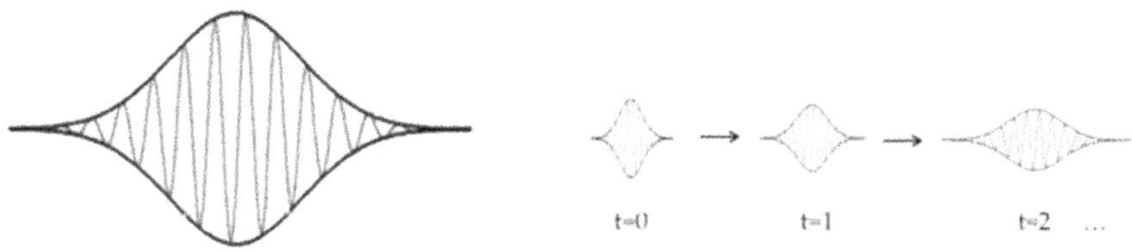

Figure 2.1 The wave packet and its spreading.

wave packets, i.e., localized superpositions of waves (see Figure 2.1, on the left). If waves of different wavelengths are present, they combine to form a localized "lump" that travels as a unit with a certain *group velocity*. If the packet's shape remains stable over time—i.e., it does not disperse—then the observed particle-like trajectories can be understood as traveling wave packets. However, Schrödinger quickly realized that this wave-only picture was not tenable, for at least three reasons. First, the *wave packet spreads:* it remains localized only for a very short time (see Figure 2.1, on the right).

Second, the equation he proposed is linear, which means its solutions obey the *superposition principle*. If the theory is universal—intended to describe everything—this leads directly to macroscopic, unobserved superpositions, rendering the theory empirically inadequate. This is the *measurement problem*, or the problem of Schrödinger's cat, which I will discuss in more detail in Chapter 3. The key issue is that if waves constitute the fundamental ontology of matter, then superpositions will be amplified from the microscopic, unobservable level to the macroscopic, observable scale. Thus, they should appear everywhere—even in contexts where we do not experience them—such as a measuring device simultaneously displaying both the result "1" and the result "2." Finally, the third problem arises in the context of many-"particle" systems: the wave function depends on the positions $r_1,\ldots,r_N$ of all the particles in the system, i.e., $\psi = \psi\left(r_1,\ldots,r_N\right)$. Therefore, it "lives" on *configuration space*—the space of all possible configurations of a system, given by the positions of all particles: $Q = \left(r_1,\ldots,r_N\right)$. Since this space has $3N$ dimensions (three per particle), and the wave function exists on this space, it becomes difficult to interpret it as a wave vibrating in ordinary three-dimensional space and propagating in time (again, see Chapter 3). As a result of these difficulties, Schrödinger arguably abandoned the idea that the wave function could serve as an ontology for matter.

Some scholars later argued that the problem with a wave ontology of matter is not the existence of macroscopic superpositions per se, but rather the fact that the different terms of the superposition can interfere—something we do not observe. If this is indeed the issue, then it suffices to identify a mechanism that prevents such interference from occurring. It has been noted that the interaction of a system with its environment can do exactly that: it can suppress interference between the different components of a superposition. This phenomenon is known as *decoherence*, as it *effectively destroys the coherence between the superposition terms* (see Joos and Zeh 2003 for more on decoherence). The effects of interference and its suppression can be represented using a mathematical device known as the *density matrix* (see the end of this section). In this formalism, interference terms appear off the diagonal of the matrix, and when these terms are suppressed, interference between the components of the superposition is effectively eliminated. However, in order for this approach to be consistent, one must accept that each term in the superposition represents a real state of affairs—even if these terms no longer interact and only one of them is ever observed. This leads to a kind of "*many-worlds*" *picture* (see next chapter): each term of the superposition evolves into a macroscopically distinct state, each representing a "world" in which that particular outcome is realized. Due to decoherence, these "worlds" do not interact with one another, but they are all simultaneously real and exist in the same spacetime. Consequently, these *worlds are transparent to one another*: the inhabitants of one world cannot interact with or perceive the objects or events in the others. This idea was likely too radical for Schrödinger, who did not pursue this line of theory further (see Chapter 7 for more).

Regardless, from a historical perspective, it became clear that a purely wave ontology for both matter and light was ultimately deemed unsatisfactory. Likewise, a purely particle ontology was also found to be untenable. One argument against it comes from the *uncertainty principle*, proposed by Heisenberg in 1927. Importantly, this is not a principle in the axiomatic sense, but rather a *theorem* that follows from the mathematics of matrix mechanics. Specifically, linear operators can have the property of *noncommutativity*, meaning that the product of two operators depends on their order: $\widehat{A}\widehat{B} \neq \widehat{B}\widehat{A}$, or $[\widehat{A}, \widehat{B}] \neq 0$. Heisenberg argued that this mathematical noncommutativity could be interpreted as implying a fundamental *uncertainty* or *complementarity*. This observation helped lay the foundations of what came to be known as the *Copenhagen interpretation* of quantum mechanics (even though members of the Copenhagen school often held differing views). According to this interpretation, properties associated with noncommuting operators cannot be measured simultaneously with arbitrary precision: the more precisely one property is known at a given instant, the less precisely the other can be known. This uncertainty relation holds generally and expresses a fundamental limit to what can be known. Most famously, it implies that one can never simultaneously measure with perfect accuracy both the position and the velocity (momentum) of a particle. This is typically written as $\left[\widehat{x}, \widehat{p}_x\right] = i\hbar$.

From this, it was inferred that even if we interpret matrix mechanics as a theory of particles, these particles cannot have trajectories. After all, for a particle to follow a trajectory, it must possess both a position and a velocity at a given instant. However, it was maintained that the uncertainty principle prohibits simultaneous knowledge of both quantities. It is important to note, though, that this inference is not logically sound: there is a distinction between what we can know—which is indeed limited by the uncertainty principle—and what actually *is*, which is independent of our epistemic access to it. Nonetheless, Heisenberg used this reasoning to argue that the concept of a trajectory is obsolete, just like the classical notions of "particle" and "wave." This marked the beginning of the (in)famous *wave-particle duality*: it was claimed that treating matter as either particles or waves is a conceptual mistake. According to this view, the question of whether physical objects are made of particles or waves is ill-posed. Instead, it was argued that quantum entities possess both localized (particle-like) and distributed (wave-like) properties simultaneously, which was taken to account for the puzzling experimental results discussed earlier. Some even proposed that because particles can be "here" and "there"—and thus "not here"—at the same time, classical logic must fail. This led to the development of *quantum logic*, a new logical system intended to accommodate such phenomena (see further readings).

Others, however—such as Einstein, de Broglie, and Schrödinger—rejected this radical approach. They advocated for a more cautious interpretation, concluding that *quantum theory, so understood, must be incomplete*. The existence of unobservable macroscopic superpositions, predicted by the theory, was taken as evidence of its descriptive inadequacy. As a result, they searched for a more complete theory—one that would include additional variables, known as *hidden variables*, which are not specified by the standard quantum formalism but would offer a fuller account of physical reality. In response to such efforts, a series of so-called *no-go theorems* were proposed, the first of which was put forward by von Neumann in 1932. These theorems purported to prove that hidden-variable theories are fundamentally impossible. However, it was later shown—first by Grete Hermann in 1932 and more definitively by John Bell in 1964—that these conclusions were not justified. Despite this, and aided by a number of historical and sociopolitical factors—such as Bohr's considerable influence within academia, the interruption of international collaboration during the war, anti-communist policies that isolated dissenting thinkers, and the natural sway of charismatic personalities—the Copenhagen interpretation promoted by Bohr gradually became the dominant orthodoxy. Over time, most physicists came to accept, as if they were unavoidable truths, the claims that realism is naïve and that physics does not describe an objective world, but merely our limited capacity to know it. The alleged justification for these claims was the repeated failure to reconcile the experimental data with classical concepts. These failures were interpreted as evidence that we lack the appropriate conceptual tools for a realist account of quantum phenomena, and that traditional notions like "particle" and "wave" are fundamentally inadequate. Instead, it was proposed that reality is bifurcated: there exists a *microscopic realm*, governed by quantum mechanics and nonclassical logic, and a *macroscopic realm*, governed by Newtonian mechanics and classical reasoning (for more on the historical development of quantum theory and its supposed inevitability, see Chapter 7).

The Fundamental Axioms of Quantum Mechanics

While some physicists resisted this attitude and pushed back against it, others were more focused on further developing the *mathematical structure* of the theory rather than engaging with its philosophical implications. Among the latter was John von

Neumann, who in 1935 provided a rigorous formulation of the *foundational axioms of standard quantum theory*. Briefly, these axioms are as follows:

1) *The complete description of any physical system is mathematically provided by the "state vector," or "quantum state."*

This axiom traces back to Schrödinger, but here it remains agnostic about whether the quantum state represents something real in the world. In other words, there is no explicit realist commitment regarding what matter is fundamentally made of. That said, the quantum state is typically represented by the Greek letters ψ or ϕ —historically, the same letters Schrödinger used. If we denote the position of ψ or N physical components of a system by $r = \left(r_1, r_2, r_3, \ldots, r_N\right)$, then the wave function is a function of all these variables: $\psi\left(r\right) = \psi\left(r_1, r_2, r_3, \ldots, r_N\right)$. (Note that, strictly speaking, this framework does not posit actual particles. So sentences like the one above should be cautiously interpreted as effective language—something like: "If r describes the position of what we have so far taken to be the N physical components of a system …") The space of all such possible configurations r is known as configuration space, as in classical physics. (Again, "configuration" might be a misnomer—there aren't necessarily any actual particles or configurations; rather, the r variables are better understood as degrees of freedom attributed to the wave function.) The space of all quantum states is a Hilbert space—a complete vector space equipped with an inner (or scalar) product. For two vectors f and g, the scalar product, denoted $< f \mid g >$ or $\left(f \cdot g\right)$, is the projection of one vector onto the other, defined as $\left\langle f/g \right\rangle \stackrel{\text{def}}{=} f g \cos\theta$, where θ is the angle between the vectors. The scalar product of a vector with itself is called its norm and corresponds to the square of its length: $< f/f > \stackrel{\text{def}}{=} \|f\|^2$. In the case of the wave function, this norm is always equal to 1, as it is tied to probabilities and must be normalized accordingly. Since the quantum state vector represents a physical system, the Hilbert space itself can be viewed as describing all possible physical states the system could be in—that is, all the ways the world could be.

A quick note on notation: Dirac's so-called *bra-ket* notation is often used to write quantum states, where a state vector ϕ is written as $|\phi>$ (a *ket*), and its dual ψ as $<\psi\,|$ (a *bra*), such that their scalar product appears as a *bracket*: $<\psi \mid \phi>$. I will avoid using this notation because it can misleadingly suggest that these objects are somehow fundamentally distinct from familiar mathematical constructs like functions or vectors—they are not.

The state vector is just that—a vector, and it can be decomposed in multiple ways, depending on the choice of basis. A basis is made up of orthogonal unit vectors that span the space. In Cartesian coordinates, for instance, the basis vectors are e_x, e_y, e_z, and a generic vector v can be written as $v = v_x e_x + v_y e_y + v_z e_z$. Likewise, in Hilbert space, one can decompose the quantum state using different bases. One particularly important choice is the *position basis*, where the wave function $\psi(r)$ arises as the representation of the state vector in terms of spatial coordinates. Because the set of all positions is uncountably infinite, instead of a discrete sum over basis vectors, one uses an integral: $\psi = \int \psi(r)dr$. (Strictly speaking, position eigenstates are not proper elements of Hilbert space, but they can be accommodated within an extension called a rigged Hilbert space, which includes generalized eigenvectors. See below for more on these.) Alternatively, one can represent the quantum state in terms of the momentum basis, which is mathematically related to the position basis via a Fourier transform of the wave function.

Now, while many prefer to refer generally to the *quantum state* or *state vector*, I tend to use the term wave function. It may seem like the former terms are broader or more precise—after all, the wave function is just one possible representation of the quantum state. However, as I'll explain in Chapter 5, I take the quantum state to be meaningful only in terms of its role in the theory, not as an object or field. From that standpoint, the choice of representation doesn't matter. But since my framework, discussed in Chapter 4, places a special emphasis on spatial position, I prefer to use the position representation of the state—that is, the wave function.

2) *If no measurement is performed, then the state vector evolves according to the Schrödinger equation.*

The Schrödinger equation can be written as $i\hbar\frac{\partial}{\partial t}\psi = H\psi$, where $\hbar = \frac{h}{2\pi}$, $h = 6.62 \times 10^{-34} m^2 kg/s$ is Planck's constant. The operator H is the Hamiltonian, which corresponds to the total energy of the system. In a common form, it is expressed as $\widehat{H} = -\frac{\hbar^2}{2m}\frac{\partial^2}{\partial r^2} + V$, where V denotes the potential energy of the system (see point 4 below). This equation is a type of wave equation, and its general solution can be written in the form $\psi = Ae^{\pm i(kr - wt)}$, as introduced in Chapter 1. Because the Schrödinger equation is linear, the superposition principle holds. That is, if ψ_1, and ψ_2 are solutions (i.e., possible physical states of the system), then any linear combination of them, like $a_1\psi_1 + a_2\psi_2$, is also a solution. Such combinations are called superposition states.

Notice how "*measurement*" *appears as a primitive term* in one of the postulates of quantum theory. This should strike us as odd: isn't measurement just another physical process—something that should itself be described and explained by the theory, rather than built into its axioms? The reason for this formulation, however, becomes clear when we consider the consequences of the superposition principle. As Schrödinger himself realized (see above), if the wave function describes everything and always evolves according to the linear Schrödinger equation, then macroscopic superpositions must exist—even though we never observe them. The logic is simple: if superpositions occur at the microscopic level and if everything (including observers and measuring devices) is described by the wave function, then these superpositions must propagate upward, to the macroscopic level. Take *the famous thought experiment involving a cat in a box,* a radioactive atom, and a vial of poison. The radioactive atom

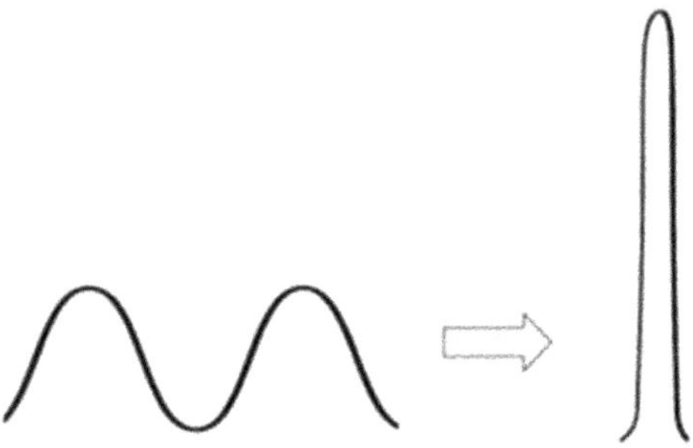

Figure 2.2 Upon measurement the wave function collapses in one of the superposition terms.

is in a quantum superposition of having decayed and not decayed. This microscopic superposition gets entangled with the fate of the cat: if the atom decays, the poison is released and the cat dies; if not, the cat remains alive. Thus, the total wave function of the system evolves into a superposition of "dead cat" and "alive cat"—a macroscopic superposition. Yet, when we open the box to observe the cat, we always see it either dead or alive—never both. This inconsistency between the theoretical prediction (superpositions at all levels) and the actual observations (definite outcomes) is what is known as the measurement problem. To deal with this, von Neumann proposed that the Schrödinger equation does not always apply. He postulated that when a measurement or observation occurs, the wave function undergoes a different kind of evolution—one that is nonlinear and nondeterministic: it "collapses" into a definite state. This led him to include a separate rule in the formal postulates of the theory: the Schrödinger equation governs the evolution of the wave function only when no measurement is taking place. Hence, he needed to specify in one of the postulates of the theory that the Schrödinger evolution was valid only unless there is a measurement. That is, the following is also the case:

> *3) When a measurement is performed, the state vector randomly and instantaneously "collapses" (or is "reduced to") one of the terms of the superposition.*

This is known as the von Neumann collapse rule. It asserts that the effect of a measurement is to induce a random and instantaneous transition in the state vector. Specifically, when a measurement is made, the wave function ψ collapses into one of the possible outcomes φ—one of the terms in the superposition. The probability of collapsing into a particular state ϕ is determined by the scalar product between ψ and ϕ, as shown in Figure 2.2.

To be more precise, the state after the measurement depends on the type of measurement being performed, as outlined in the following:

> *4) Observables (i.e., the measurable properties of a physical system) are represented by linear self-adjoint (or Hermitian) operators. The possible values of an observable, associated with a given operator, are determined by the eigenvalues of that operator. When a measurement is performed, the state vector collapses into the eigenstate corresponding to the observed eigenvalue.*

Such a rule is also called the eigenvalue-eigenstate link, and it associates measurable quantities (energy, position, velocity, and the like) with a particular type of operator. As anticipated, operators are often represented with a "hat" on top, as in $\widehat{A}$. For example, the momentum operator is $\widehat{p} = -i\hbar\frac{\partial}{\partial r}$, so the operator corresponding to the total energy $E = \frac{p^2}{2m} + V$ is given by the Hamiltonian operator $\widehat{H} = -\frac{\hbar^2}{2m}\frac{\partial^2}{\partial r^2} + V$. While the evolution of the state vector comes from Schrödinger's work, this postulate instead comes from Heisenberg's matrix mechanics.

An "eigenvector" ϕ_a of an operator $\widehat{A}$ is a vector such that, when the operator acts on it, it provides as result the same vector ϕ_a multiplied by its "eigenvalue," $\alpha : \widehat{A}\phi_a = \alpha\phi_a$. A self-adjoint operator is defined as one with eigenvalues which are real numbers. Self-adjoint operator is defined as one with eigenvalues that are real numbers. The set of an operator's eigenvalues is called its spectrum. By definition, the spectrum of operator $\widehat{A}$ provides the set of possible measurement results for the quantity $\widehat{A}$ represented by $\widehat{A}$. If one is performing a measurement of the property represented by $\widehat{A}$, the state vector can be seen as decomposed into a superposition of eigenvectors ϕ_a of $\widehat{A}$, weighted by their respective eigenvalues $\alpha_i : \psi = \alpha_1\phi_{\alpha_1} + \alpha_2\phi_{\alpha_2} + \ldots + \alpha_N\phi_{\alpha_N} = \sum_i \alpha_i\phi_{a_i}$. If the result of a measurement of A is α_i, then the state vector collapses into the corresponding eigenvector ϕ_{a_i}. For instance, when the Hamiltonian H does not depend on time, the Schrödinger equation is written as $H\psi = E\psi$, or equivalently $-\frac{\hbar^2}{2m}\frac{\partial^2}{\partial r^2}\psi = E\psi$. This is an eigenvalue equation for the H operator, and it gives as results the possible experimental values E of the energy (which, depending on the circumstances, may be continuous or discrete).

5) *During the measurement of a quantity corresponding to some operator, the probability of finding a particular eigenvalue at a given time is determined by the scalar product of the initial state and the eigenvector corresponding to that eigenvalue.*

This is the Born rule, which governs the statistics of experimental results. In formula form, $Prob(\alpha_i, t)$, the probability of finding a particular result α_i at time t, is given by the square of the scalar product of ψ and $\phi_{a_i} : Prob(\alpha_i, t) \overset{def}{=} \left\|\langle \psi / \phi_{a_i} \rangle\right\|^2$. It is important to note that this postulate makes standard quantum mechanics a fundamentally indeterministic theory, in contrast to classical mechanics, which is deterministic.

Additionally, since the probabilities depend on the square of the scalar product of two wave functions, if two wave functions differ only by a phase factor—for example, ψ_1 and $\psi_2 = \psi_1 e^{i\theta}$ —they will have the same scalar product and therefore yield the same empirical distributions. This fact is often summarized by stating that *the wave function is a "ray" in Hilbert space*, where a ray refers to the equivalence class of wave functions that differ only by a phase factor.

Composite Systems, Entanglement, Nonseparability, and Nonlocality

It turns out that a *composite system*, made up of several interacting subsystems, may be described by very peculiar states. To illustrate this, any state could be used, but it is typically done using spin states. As we mentioned earlier, the discovery of new data led scientists to postulate the existence of a new property, which is considered to be exclusively quantum. This property was called "spin" because, initially, it seemed as though the particle would rotate around its own axis. Later, it was realized that this is just a metaphor, as no actual rotation occurs, but the name remained. The spin of a particle can be assigned to any direction—x, y, or z—and it is quantized, meaning it does not take continuous values. Depending on the type of particle, the eigenvalues of the spin operator along a given direction can be either integer or semi-integer. For particles with semi-integer spin, the values are $+1/2$ or $-1/2$, which are also called "up" and "down," or sometimes represented as $+$ and Integer spin applies to particles known as bosons, like photons, while semi-integer spin applies to particles called fermions, like electrons and protons. For fermions, σ_x^+ denotes the spin state along the x-direction being "up," with similar notations for "down" and for the other directions.

If one has a two-particle composite system with semi-integer spin, its possible states include ones like $\sigma_x^+\sigma_x^-$, representing the first particle being "up" and the second being "down" along the x-axis. One can interpret these *product states* as describing two particles: the first with spin "up" and the second with spin "down" along the x -direction. (These states are called "product states" because the state of the composite system is literally the product of the states of the two individual particles. A second subscript could be used to identify the particles, like $\sigma_{x1}^+\sigma_{x2}^-$, but this notation is often avoided for simplicity. Instead, it's assumed that the order of terms matters: the first term refers to the first particle and the second to the second particle.) However, other types of states are also possible, such as the following state called the singlet state: $\frac{1}{\sqrt{2}}\left(\sigma_x^+\sigma_x^- - \sigma_x^-\sigma_x^+\right)$ (The numerical factor in front ensures that the state is normalized to 1, which is required by the connection between the norm of the wave function $\|\psi\|^2$ and probabilities, as *per* the axioms of quantum mechanics.) While the interpretation of product states is straightforward, the singlet state, which is an example of an *entangled state, does not assign definite spin properties to each particle.* Instead, the particles exist in an indistinguishable mixture: neither particle has a definite spin along the x-axis. Thus, they are said to be *nonseparable*; in other words, "the whole is not the sum of the parts" (Howard 1985). The properties of each individual particle do not determine the properties of the composite system because there are no definite properties for the individual particles—only properties of the whole system. An important feature of entangled states is that, because they are so deeply interconnected, it is possible to affect one particle by interacting with the other, regardless of how far apart they are. That is, *their components can interact nonlocally.* For example, if the two particles travel in opposite directions, one might end up in Chicago, Illinois, and the other in Honolulu, Hawaii. By interacting with the particle in Chicago, one can instantaneously affect the particle in Honolulu. In other words, these particles interact in a nonlocal manner: what happens to one particle in a given spatial region instantaneously affects what happens to the other, regardless of the distance between them. While nonseparability is a feature of quantum systems, nonlocality describes a characteristic of the interaction itself. This "*spooky*" *action at a distance*, inherent to standard quantum theory, is at the heart of the objections raised by Einstein, Podolsky, and Rosen (1935). They argued that, if the spin of a particle is measured along one direction and found to be "up," then the spin of the other particle is instantaneously determined to be "down," even though, as we have seen, quantum theory does not assign any definite spin value to each particle before the measurement. In other words, as a result of being in an entangled state, the act of measuring one particle's spin causes the state of both particles to collapse simultaneously. The measurement on the Chicago particle instantaneously collapses the state of the Honolulu particle, even if they are far apart.

This nonlocal action (in the sense that what happens in one place may instantaneously influence what happens in another, arbitrarily distant place) was considered by EPR to be *unacceptable* for various reasons, both scientific and intuitive, as discussed in Chapter 1. For one, relativity theory dictates that there is a finite maximum velocity—the speed of light. In classical physics, forces also act instantaneously. However, as we saw in the case of electrodynamics, the concept of the electromagnetic field, an entity that spreads out everywhere and travels at the speed of light, allows us to view the interaction as local, mediated by these fields. In contrast, no such mediator exists in quantum theory. Additionally, the fact that causes seem to precede their effects suggests that the interaction between objects is not instantaneous but occurs at a finite velocity. This principle is central to the concept of locality, often referred to as local causality. Furthermore, we commonly describe systems as if they are isolated because we assume that the influence of distant objects on their motion can be ignored. We break the world down into local parts and study them individually, in isolation. However, if interactions were instantaneous, we could no longer think in these terms. In classical mechanics, the intensity of forces quickly decreases with distance due to the $1/r^2$ dependence of the force, making them negligible under certain approximations—such as when an object is far from others. This does not hold true in the case of quantum correlations, whose strength is independent of the distance between the two particles composing the system. Therefore, while it is often said that a violation of locality conflicts with relativity, it seems to me that it is also in tension with our general understanding of physics (for more on this, see Chapter 4).

Consequently, EPR concluded that the only remaining explanation for the correlations between the two particles (where every time one is found to be "up," the other is "down") is to *assume that standard quantum theory is incomplete*. In this view, each particle must have had definite spin properties all along, even though standard quantum mechanics claims otherwise. According to EPR, the measurements simply reveal these pre-existing properties. That is, the observed correlation in experimental results can be explained (locally) by the correlations the particles had from the start. In other words, the measuring devices are revealing a *pre-existing* value of the spin property of each particle. EPR argued that standard quantum mechanics is incomplete because it does not assign pre-existing spin properties to particles in entangled states. These spin values are "hidden variables" that are necessary to supplement the standard quantum theory, enabling an explanation of the correlations in the results without abandoning locality.

Several years later, Bell (1964, 1966) sought to explore the consequences of completing quantum theory in the way that EPR proposed—i.e., completing it while preserving locality. This led him to derive a specific inequality, a mathematical relationship that should hold true if such a "completed" quantum theory were valid. Since standard quantum mechanics predicts a violation of this inequality, it could serve as a crucial test to distinguish between the two theories: an experiment could be conducted to check whether the inequality holds or not. If it holds, then quantum mechanics would (arguably) be falsified; if it is violated, quantum mechanics would be confirmed. Subsequent experiments (which later earned the Nobel Prize in Physics) convincingly showed that quantum mechanical predictions match the experimental data (Aspect et al. 1982). This implies that *no completion of quantum theory aimed at explaining quantum correlations locally is empirically viable*. The argument goes as follows:

- EPR: Given the quantum predictions, since locality is true, *there must be pre-existing values of spin properties* that quantum mechanics does not postulate.
- Bell: Completing quantum mechanics with these pre-existing spin properties results in *Bell's inequality*, which is violated by "incomplete" quantum mechanics.
- Aspect: Experiments show a *violation* of Bell's inequality.
- Conclusion: *The EPR locality assumption is false.*

Bell's inequality arises from the assumption that the observed perfect correlations must have a local explanation, grounded in the properties of the particles. Since Bell's inequality can be empirically tested and is found to be violated, the perfect anticorrelations cannot be locally explained. In other words, only nonlocality can account for the quantum correlations. Bell believed he had proven this, as locality was the only assumption made by EPR. In Chapter 4, I will discuss the pilot-wave theory, which can be considered a hidden variable theory but is not affected by this argument. In this theory, hidden variables are not used to locally explain the perfect correlations, which are explained by nonlocality. Rather, the hidden variables ground the theory in spacetime, specifying the fundamental ontology of matter.

To conclude, it's important to note that there is another assumption in Bell's proof that is rarely challenged: *statistical independence*. This assumption states that the experimental results are independent of the choice of experiment to be performed. Technically, one could avoid the nonlocality conclusion by rejecting statistical independence (see also Chapter 7). This leads to theories known as "*superdeterministic*," which Bell himself deemed untenable. These theories deny statistical independence and assume that the outcome of an experiment is not determined by the properties that the measured system actually possesses, but rather by the measurement choice. Alternatively, one could avoid the nonlocality conclusion by allowing for *backward causation*: the result for one system is not determined by a nonlocal interaction with the other system, but rather by the result being "communicated" into the past system to adjust itself accordingly. It is beyond the scope of this book to discuss these theories in detail (see Chapters 4 and 7 for further exploration). At this point, however, I am inclined to think that resisting the nonlocality conclusion by rejecting statistical independence is suspiciously reminiscent of a Ptolemaic view, where one denies that the telescope works when observing the phases of Venus, rather than a Copernican view, which embraces the anomalies in the orbits of Uranus.

Density Matrix and Reduced Density Operator

There are alternative ways to describe systems besides using the wave function. One such approach is the use of the *density operator*. A given (nonentangled) wave function, $\psi(r_1,\ldots,r_N)$, is associated with the following matrix: $\rho\left(r_1,\ldots,r_N,r_1',\ldots,r_N'\right)\stackrel{\text{def}}{=}\psi\left(r_1,\ldots,r_N\right)\psi\left(r_1',\ldots,r_N'\right)$. The operator corresponding to this matrix in the position representation is called the density operator $\widehat{p}$ and it is defined as the operator that projects onto the state ψ. In Dirac notation, this is written as $\widehat{p}\stackrel{\text{def}}{=}|\psi\rangle\langle\psi|$, so that formally we have $\widehat{p}|\psi\rangle=|\psi>$.) Since it can be shown that the expected measurement results for a given operator $\widehat{A}$ are given by the sum (or integral) of the diagonal elements of the density matrix when applied to $\widehat{A}$ (this is called the trace of $\widehat{\varrho}\,\widehat{p}$, $Tr(\widehat{\varrho}\widehat{A})$), then the density matrix contains the same information of the system as the wave function.

If a system is part of a larger system, such as an electron in a multi-electron system or an entangled state, there is no individual wave function for it; the wave function describes the composite system as a whole. However, if the system is in a *mixed state* (where the individual system may exist in a variety of states), the density operator can be generalized as the sum over all projectors onto the possible states, weighted by their respective probabilities. That is: $\widehat{p}\stackrel{\text{def}}{=}\sum_i|\psi_i><\psi_i|$. In this way, the state ϱ_A of a subsystem A (e.g., electron 1) in a composite system of A and B (e.g., electron 1 and electron 2) can be written as the trace over subsystem B of the density matrix of the composite ϱ_{AB}: $\varrho_A=Tr\,\varrho_{AB}$. For instance, for the entangled state of two electrons with opposite $x-$ spin, we have the composite wave function: $\psi_{1,2}=\frac{1}{\sqrt{2}}(\sigma^+\sigma^- -\sigma^-\sigma^+)$ (suppressing the x subscript in σ, for convenience). The density matrix is $\varrho_{1,2} = \begin{bmatrix} 1/2 & 1/2 \\ 1/2 & 1/2 \end{bmatrix}$. The reduced density matrix describing electron 1 is $\varrho_{1,2} = \begin{bmatrix} 1/2 & 0 \\ 0 & 1/2 \end{bmatrix}$.

Interestingly, while the wave function is defined up to a phase, the density matrix is unique. This has led some to suggest that one should focus on the density matrix rather than the wave function as the fundamental ontology of matter (see Chapter 3).

The Challenges from Quantum Theory and the Rise of Structuralism

In the first section of this chapter, I briefly reviewed the history of quantum theory to understand why many came to believe it was incompatible with realism. New empirical data suggested contradictory ontologies, and our repeated failures to explain them led some to adopt anti-realism or at least to set aside metaphysical questions in favor of developing a set of rules to systematize experimental results. This is essentially what von Neumann did. As we saw in the previous section, he introduced the notion of measurement directly into the axioms of the theory. From a perspective in which quantum theory is viewed as fundamental and complete, this is problematic: Why shouldn't the theory also apply to measurement processes? However, the urgency of this issue diminishes if one's only concern is accurately reproducing experimental outcomes. Practically speaking, it's perfectly clear what an experiment or measurement is: you go into a lab, make objects interact, and observe what results. In other words, if one gives up on the idea that physics must offer a faithful depiction of reality and instead adopts a more pragmatic view—that physics merely predicts what a scientist should expect from an experiment—then the von Neumann formulation is entirely adequate. Interestingly, some have argued that it is possible to accept von Neumann's collapse rule as it stands and still maintain a realist stance. Let us explore how that might be possible in the next section.

Macroscopic Ontology

A realist believes that a fundamental physical theory is (approximately) true, not just useful. Proposals like the *information-theoretical (IT) approach* (Bub and Pitowsky 2010) suggest that one could make sense of standard quantum theory in a realist way by assuming that measurements are fundamental and not further analyzable. In line with the original ideas of the Copenhagen school, as I described earlier, this approach proposes a *macroscopic ontology of measurement outcomes*. In this framework, there is no issue in deciding what constitutes a measurement in order to determine when to apply the collapse rule. Measurements are fundamental, and they are not understood in terms of interactions between microscopic entities. Moreover, this approach requires a shift in thinking: one can no longer view electrons and protons as composing macroscopic objects, nor can one view quantum theory as providing a constructive explanation. Bub and Pitowsky explicitly defend the idea that quantum theory is a *principle theory*: the Hilbert space formalism, as outlined earlier, imposes constraints on what can and cannot physically occur, similar to how the principles of thermodynamics impose constraints. In this framework, the wave function does not represent matter or any other feature of nature. Instead, it describes our ignorance about the actual state of the system. In other words, *the wave function is epistemic*, not ontic. However, this approach remains objective in the sense that measurement outcomes are objective, and the constraints imposed by the Hilbert space structure are also objective. Proponents of this view argue that it is compatible with realism, as quantum theory provides an objective description of what we experience and tells us what to expect when performing a measurement according to the quantum rules.

A similar approach is *Quantum Bayesianism, or Qbism* (pronounced "cubism"). Qbists argue that quantum theory aims to describe reality beyond mere appearances, though they acknowledge that we still do not know what the ontology is (Fuchs et al. 2014). Related proposals, such as those by Healey (2020), take a *pragmatist approach*, treating the wave function as a guide to what we can expect in experimental situations. We will discuss these approaches further in Chapter 5.

Structure-Oriented Metaphysics

Another proposed way of making standard quantum theory "as is" compatible with realism is through *structuralism*, which was briefly mentioned in Chapter 1. As anticipated, while epistemic structuralism (which holds that we are justified in believing in the existence of structure rather than objects) was introduced as a response to the pessimistic meta-induction argument against realism in general (Worral 1989), ontic structuralism was (at least partially) motivated by the challenges to realism posed by quantum theory. These challenges are not tied to the idea of measurement being fundamental. In fact, structuralists argue that their approach dissolves the measurement problem (see Chapter 3). Instead, the motivation for ontic structuralism stems from concerns about individuality and indistinguishability, which are central to quantum theory, as discussed in the next section.

The core idea behind ontic structuralism is that, contrary to traditional approaches where objects are considered fundamental, *structure itself is fundamental*. There are different variations of ontic structuralism, with the most radical being *eliminativism*, also known as radical ontic structuralism (French 1998, Ladyman 1998, French and Ladyman 2003, 2011). According to eliminativism, quantum theory fundamentally deals with structural relations. The key tenet is that there are *no fundamental objects as individuals*. Instead, objects emerge as useful, contingent patterns (see the end of this section). In this view, there are only relations and purely relational structures: fundamental physical relations do not require fundamental objects as *relata*. This idea is sometimes encapsulated in slogans such as "there are only relations, without *relata*" and "it's relations and structures all the way down."

In addition to the specific motivations arising from quantum theory, structuralism is generally appealing due to its *ontological parsimony and simplicity of description*: Why postulate two categories (objects and relations) when one (relations) suffices? Furthermore, structural descriptions are *fruitful*, as they apply in many different contexts. Structuralism also carries the virtue of *humility*: due to concerns about underdetermination, we lack sufficient evidence to postulate the existence of objects. As a result, structuralists advocate for a more modest approach that avoids assuming the existence of objects unnecessarily.

Notice that structuralists bypass the question of the nature of the wave function. They do not consider the wave function to be a property or a "substance," as is proposed by wave function realists (see the next chapter). Instead, it is viewed as part of the structure constituted by the laws and symmetry principles of our fundamental theories. In this sense, within structuralism, the Schrödinger equation should not be understood as governing the evolution of an object (the wave function). Rather, it expresses "the dynamic nature of the structure itself." As French (2014) puts it, for the structuralist, laws represent the fundamental nature of the world. Eliminativism contrasts with a more moderate version of ontic structuralism (introduced in Chapter 1 and discussed further in Chapter 5), according to which objects are as fundamental as the structural relations between them. Other proposals that share similarities with structuralism include *relational quantum mechanics* (Rovelli 1996), where objects arguably emerge from a purely relational structure. However, relational quantum mechanics seems to have different motivations from structuralism, as it can be interpreted as a refinement of the standard view, in which any system can be seen as an observer and any interaction as a measurement (Laudisa and Rovelli 2021).

Arguments for Structuralism

If we focus on quantum mechanics, there are *three main arguments* for eliminativism, and more generally for structuralism: the argument from entanglement, the metaphysical underdetermination argument from quantum statistics, and the argument from Humean supervenience.

The basic idea behind *the argument from entanglement* is that entangled particles are individuated by their structural relations, given that they lack individual properties. For example, two entangled particles in a singlet spin state are entirely described by their singlet state wave function. As we've seen, the standard theory does not assign properties to either individual particle. The state has a property of being in a total spin 0, but it does not specify the spin property of each particle. This consideration has led some to adopt a structuralist viewpoint: *there is no property that individuates each particle; rather, they are fundamentally described in terms of their relations to each other.* For a particle, being in a given state means being in a particular relation with the other particle, and that's all.

A second argument uses *quantum statistics to show that it is underdetermined whether particles have individuality, so we are not justified in believing they do.* In classical physics, two particles are considered identical if they belong to the same species and have the same individuating properties, such as mass and charge—both could be electrons, for example. This concept has nothing quantum about it, as one can talk about identical particles in classical mechanics. Even if particles are identical, classical particles can often be distinguished by their location: one electron is here, the other is there, so they are distinct and can be labeled. Since classical identical particles can be labeled, they obey a particular statistic. When considering a set of identical particles, it is often impractical, or even impossible, to know their exact arrangement. So, it is useful to count the possible ways particles could be arranged and use statistical considerations to establish the probability of a given distribution. Consider two identical particles, labeled 1 and 2, and their possible arrangements into two boxes, left and right. These arrangements are: (1) both particles in the left box, (2) both particles in the right box, or (3) one particle in the left and one in the right. Notice that the third configuration counts (or weights) twice as much as the other two, in the sense that swapping particle 1 with particle 2 in the arrangement (3) gives the same distribution. Therefore, the probability of having the particles in different boxes is 1/2. There are four possible distributions, and distribution (3) accounts for two of them, so the probability is 2/4 = 1/2. This is the *Maxwell-Boltzmann distribution*, which was used to reduce thermodynamics to Newtonian mechanics, assuming particles can be labeled. In standard quantum mechanics, however, the situation is different. Quantum particles—whatever we mean by "particles" in this theory—can be identical if they belong to the same species, but they can also be indistinguishable. In fact, they may have the same properties, including location, since they are entirely described by the wave function. Moreover, in standard quantum theory, all that matters are the experimental outcomes. It doesn't matter which particle plays which role to produce a given measurement result. Therefore, the indistinguishability postulate holds: if we exchange one particle with another, nothing changes in the values of the quantum observables. Since indistinguishable particles cannot be labeled, Maxwell-Boltzmann statistics no longer apply. The third configuration is realized in only one way, just like the first two. From the indistinguishability postulate, it follows that the wave function transforms under permutation (label reshuffling) in only two possible ways: it either remains unchanged (symmetric) or changes sign (antisymmetric). These correspond to bosons and fermions, respectively. The Pauli exclusion principle, which states that no two fermions (such as electrons) can occupy the same quantum state, is a special case of this. Thus, there are two quantum statistics: *Bose-Einstein statistics* for bosons and *Fermi-Dirac statistics* for fermions. In the former, particles can be arranged in the same three configurations as in Maxwell-Boltzmann statistics: (1) both in the left box, (2) both in the right box, and (3) one in the left and one in the right. However, as noted earlier, configuration (3) counts equally with configurations (1) and (2), so the probability of one particle in each box is 1/3. In the case of fermions, the probability is 1, as the Pauli principle dictates that only one particle can be in each box. Now, how should we understand the nature of these particles? Do these identical, indistinguishable particles possess any sort of individuality? Arguably, an object is an individual if it has a property that other objects lack. So, if an object has no individuating property, it is not an individual. We might think of *quantum particles as individuals if we can distinguish them*, as we can with classical particles. This could be done in terms of their "haecceity"—their primitive "thisness." However, haecceity is a super-empirical, nonqualitative, or transcendental feature, and therefore can't be supported by empirical evidence. Alternatively, we might think of *quantum particles as objects that are not individuals*. This more radical view seems to fit better with quantum statistics, though it is certainly more controversial. Nonetheless, it is argued that, in the absence of a decisive argument, the situation is *underdetermined*, and both views—particles as individuals and particles as nonindividuals—appear equally supported (French and Redhead 1988, French and Krause 2006). Indeed, a key principle underlying structuralism is that we are not justified in postulating the existence of something whose existence cannot be known through purely empirical means. These considerations lead structuralists to view all cases of underdetermination as evidence for their perspective. The idea, then, is that quantum statistics suggest it is underdetermined whether quantum objects are individuals. From this, eliminativists conclude that *there are no objects, only structure*. Similar arguments are made based on symmetries, invariances, and the existence of multiple empirically equivalent mathematical formulations of the same theory (see "Further Readings").

A final argument for structuralism concerns the *failure of Humean supervenience*. There is a prominent conception of the laws of nature, known as the Humean view, according to which laws "supervene" (i.e., they are determined by) the total state of affairs of matter. According to this view, there is nothing beyond the Humean mosaic, which consists of a local collection of fundamental entities and their properties. This is the thesis of Humean supervenience. It has been argued that Humean supervenience fails if quantum theory is true (Teller 1989). The reason for this is that *the wave function is a nonlocal object*, as it is defined in configuration space, while the Humean mosaic is said to be local. In light of this, structuralists argue that it is better to *abandon the idea that nonlocal relations are instantiated by individuals* (as some Humeans propose) and instead adopt an ontology of fundamental structural relations that are not grounded in objects (French 2014).

The Emergence of Objects

I have briefly surveyed the motivations for endorsing a structuralist perspective, specifically arising from quantum theory. The view provides a clear ontology: structure. However, one might still find this perspective unsatisfying. In addition to objections specific to structuralism, which we will discuss in the next section, there are other, more general concerns. First, this theory begins with standard quantum theory, in which the notion of measurement is embedded in its very formulation. Like the IT approach, structuralism seeks to account for measurement results. This reflects the empiricist spirit that guides structuralism: the idea is to engage in naturalized metaphysics with a strong empiricist foundation. The underlying philosophy is to remain humble and refrain from speculating beyond what is empirically given. As a result, the type of realism proposed by structuralism is relatively weak, bordering on instrumentalism.

That said, structuralists propose that *objects, both at the microscopic and macroscopic levels, emerge from structure* (Ladyman and Ross 2007). This means that macroscopic entities are not composed of microscopic constituents, and, as a result, compositionality and constructive explanations are impossible. Instead, both microscopic and macroscopic objects are seen as *real patterns,* defined by their usefulness (see also Dennett 1991, Saunders 1993, 1995, Wallace 2003, 2012). What we call "particles" are neither fundamental nor composed entities. Rather, they are merely *useful fictions* that allow us to conveniently express certain regularities at a given level of description. The same holds true for chemical compounds, molecules, tables, chairs, and measurement devices. These entities are not analyzed in terms of more fundamental components because such components do not exist. Instead, they are understood as effective descriptions. This position, known as *rainforest realism*, is a realist position: theories describe the world, and while fundamentally only structure exists, many types of entities emerge at various levels. This approach aligns with the idea that a theory explains phenomena in terms of principles rather than through a dynamic process. In fact, since compositionality fails, *the explanation provided by structuralism is not bottom-up but fundamentally top-down, principle-like* (see also the final section of this chapter).

Objections against Structuralism

Structuralism has garnered significant attention in the philosophy of physics, both in support of it and in opposition. As we saw in Chapter 1, structuralism originally emerged as a response to the pessimistic meta-induction argument, but it later gained prominence as the most parsimonious, humble, fruitful, and natural ontology for quantum theory (Ladyman and Ross 2007). These characteristics are highly appealing to many philosophers engaging in naturalized metaphysics for various reasons. However, while this view continues to captivate many, it has also faced considerable criticism from others.

The Need for Fundamental Objects

The most obvious objection to structuralism is that it is not intelligible. Structuralists, particularly eliminativists, argue that the fundamental ontology is structure. But *what exactly is structure?* If we understand structure as a type of relation, the natural follow-up question is: A relation between what? *This view seems incoherent*, as according to eliminativism, there are only relations, with objects emerging as needed. However, relations cannot exist without their *relata*—i.e., the objects that structures are relating—also existing at the fundamental level (Chakravartty 1998, Psillos 2001).

Other objections related to the nonfundamentality of objects in eliminativism concern properties, causation, and laws. These concerns are grounded in the belief that *objects are necessary to ground properties, that properties are essential to explain causation, and that causation is tied to laws.* For instance, structuralism seems unable to account for intrinsic properties like mass and charge, as properties require objects (Ainsworth 2010). In other words, structuralism argues there are no intrinsic properties of objects because there are no objects at the fundamental level, which many consider a problem. Similarly, it has been argued that structuralism cannot explain causation because objects with appropriate properties are essential for causation, which itself requires two distinct objects—one causing the other (Busch 2003, Chakravartty 2003, Psillos 2006).

Moreover, many critics claim that *the relationship between structure and laws is insufficiently detailed*. Structuralist laws don't seem to align well with our traditional conceptions of laws. There are two main views about laws. One is the Humean view, mentioned earlier, which asserts that laws "supervene" on the arrangement of instantaneous, point-sized instantiations of "perfectly natural" qualities: the Humean mosaic. According to this view, laws can be derived directly from the mosaic. In contrast, anti-Humean views reject Humean supervenience, often positing that laws exist independently of the mosaic and govern the behavior of objects. Alternatively, some anti-Humeans focus on dispositional properties, claiming they are responsible for the nomological power we typically associate with laws. For example, in the classical picture of a ball falling in a gravitational field, a Humean would say that the ball falls because of the properties in the Humean mosaic—there's no gravitational law causing the ball to fall. Instead, the gravitational law is a convenient summary of the situation we observe. By contrast, anti-Humeans would say that the gravitational law itself is responsible for the ball's behavior, while dispositionalists argue that properties like mass are what make the ball fall under specific circumstances. Structuralist laws, however, don't supervene on the Humean mosaic, don't govern, and have no properties of their own. This raises the objection that *it's unclear what laws are in structuralism* and how the structuralist position fits into the broader debate about the laws of nature.

Additionally, it is argued that *structuralism fails to sufficiently explain the observed correlations between objects and their properties*. For instance, it's an observable fact that certain fundamental particles have specific properties, like mass and charge. There's a correlation between a particle's mass and charge, e.g., each electron has a specific mass and charge. Traditionally, this is explained by saying that an electron is an object, and these properties define that object. Structuralism, however, cannot provide this answer, as it denies the existence of fundamental objects.

A variety of replies have been proposed to address these objections. To the charge of unintelligibility, several accounts have been suggested to define "relations without *relata*." For instance, one could postulate that the fundamental structure is primitive and unanalyzable. Other replies are possible, though whether they succeed is a matter of controversy. Alternatively, one could adopt a more *moderate version of structuralism*, in which both objects and structure exist at the fundamental level (Esfeld 2004, see later).

When responding to objections about properties, laws, and causation, structuralists mostly "*bite the bullet*." They argue that properties and laws can emerge from structure, just as objects do. For example, it has been suggested that properties relate to symmetries or are linked to suitable group representations (Castellani 1998, French

2014). Alternatively, one could deny that laws require properties, proposing that laws, as part of the structural relations, are fundamental and can be defined as "whatever determines the behavior of such structure." Similarly, for the objection concerning the bundle of properties, structuralists could argue that structure, rather than objects, bears the *properties as emergent features* (French 2014). As for causation, one could either maintain that physics has no place for causation (Ladyman 2008) or propose an account of causation in terms of structure (French 2006, Esfeld 2009).

The Vagueness of Structure

Other objections are more concerned with the notion of structure itself, rather than the fundamentality of objects. One criticism is that structuralism relies on *a notion of structure that is too vague.* Critics argue that structuralism should first define what structure precisely means and then link it to the physical content of the theory. Another related objection is that structuralism *conflates mathematical and physical structure.* Structuralists often urge us to focus on the equations of the theory, suggesting that this is the structure we should be realists about. However, these equations express mathematical relations, not physical ones. So, how can we distinguish between mathematical and physical structure within structuralism? It seems that physical structure is simply mathematical structure instantiated in objects, but since structuralism denies the existence of objects, a different definition of structure is needed (van Fraassen 2006). Along similar lines, it has been argued that *a realist must distinguish between form and content,* but this distinction seems impossible within structuralism. A realist, in this context, is someone who claims that what is represented by the mathematical expressions of a theory is true, in the sense that it refers to something in the world. The form is the equation, while the content is what the equation refers to. However, in structuralism, there seems to be no referent—no content—beyond the form itself. The only way structuralism could be viable is if it assumes such a distinction exists. If this is the case, some argue that structuralism would collapse into a more conventional form of realism (Psillos 1995).

Structuralists might respond by saying that what distinguishes mathematical from physical structure is not that the latter is instantiated by objects, but that it is *instantiated by the structure itself.* Furthermore, it is not necessarily the case that structuralism collapses into standard realism. One could think of the content of the theory as the physical structure, while the form is the mathematical one. Saunders (2003) argues that there is no reason why structuralism must insist that structure is mathematical, while Tegmark (2007) embraces a form of Pythagorean structuralism.

Perhaps the more serious worry is this: *structuralism does not resolve the pessimistic meta-induction argument*, as structure is often lost during theory change (Chakravartty 2004, Newman 2010). In particular, it has been argued that structuralism fails in the transition from classical to quantum theory. Structuralists claim that the wave function represents the structure, but there is no classical analog for it. So, how can structure be preserved in the classical-to-quantum transition (Allori 2018a)?

The Costs of Structuralism

Some of the replies to the objections above imply adopting *a radical metaphysics*, which requires revising our commonsensical concepts. Regardless, the view is coherent. The question then, as always when considering radical changes, is whether its advantages outweigh the costs. As many before me have argued (Psillos 2001, Chakravartty 2004, Morganti 2011), I believe the benefits do not outweigh the problems.

The aspect of structuralism that I find most challenging is the type of explanation it provides, which constrains the kind of realism this framework can offer. Structuralists explicitly endorse a *top-down, principle-based explanatory framework*, which is at odds with a constructive approach. They need this framework because they use principles to solve the measurement problem, as we have seen: measurement devices, like all objects, are emergent useful patterns. In the language used in Chapter 1, they provide a principle theory. However, compared to a constructive explanation, the structuralist approach only explains *what to expect* in an experimental situation, *without offering a reason for why* the expected phenomenon occurs. This approach is realist, but its explanatory power is weaker than that of constructive approaches. This is why some argue that this view is closer to empiricism than a committed realist would prefer (Egg 2019). In this sense, the structuralist view is similar to the IT approach to quantum theory, discussed earlier in this section, according to which quantum theory is understood as laying out a set of constraints imposed on empirical data. These views are realist about the fact that these data exist objectively and mind-independently, but they do not think there is a need to tell any additional story about how these data are generated from a microscopic perspective. Structuralists disagree with the IT approach, particularly its claim that measurements are fundamental, as this contradicts the spirit of structuralism. For structuralists, measurement outcomes cannot be part of the fundamental ontology of the world because only structure is fundamental. As discussed, structuralists take objects and processes (like measurements) to be emergent patterns. They regard these, both at the macroscopic and microscopic levels, as useful fictions—convenient tools for describing phenomena at a given level, but not things that should be understood as actually existing as composite objects. To claim that measurement devices are not made up of particles is an explicit rejection of constructive explanation.

In other words, in the constructive view, macroscopic properties arise from the dynamics of microscopic entities. This is not the case with structuralism or other principle-based explanations. *The explanations in these theories are nondynamical*: they do not involve the mutual interaction of systems. I believe that these two different types of explanations—constructive and principle-based—reflect different attitudes toward realism. Specifically, the type of explanation one favors is linked to the type of realism one endorses. A realist who favors constructive explanation is committed to the existence of microscopic entities that compose macroscopic objects and whose motion and interactions account for macroscopic phenomena. In contrast, those who favor IT or structuralism, which are principle-based, do not require this commitment. They can be realist about macroscopic phenomena without postulating any microscopic entities. Principle-based explanations "remain on the surface" of macroscopic observable phenomena, without "going deeper" to postulate unobservable microscopic entities whose existence cannot be straightforwardly justified. This approach seems to align with the fact that structuralists take the underdetermination of theories by data very seriously. For someone who favors constructive theories, it would be unproblematic to break any underdetermination using considerations involving unobservable entities, while structuralists would avoid doing this due to their empiricist stance. Therefore, the type of realism endorsed by those favoring constructive explanations may be dubbed robust, while principle theorists like structuralists, due to their closer alignment with empiricism, endorse a weaker form of realism (Allori 2023a).

Evaluation

To summarize, structuralists maintain that structure is fundamental. Objects, both at the microscopic and macroscopic levels, emerge from this structure. Laws are understood as structural connections, and there are no properties or causal powers at the fundamental level. As we've seen, the main advantages of structuralism are its simplicity, fruitfulness, and empirical humility. However, its main cost, from my perspective, is that it is revisionary both in terms of ontology and the type of explanation it provides. This results in a *weak form of realism based on principle explanations.*

As we've noted, the plausibility of a radical view always depends on a cost-benefit calculation. The fact that there are *less-radical alternatives* offering similar benefits seems to decrease the plausibility of structuralism. For instance, moderate versions of structuralism have been proposed partly in response to some of the issues discussed here. Contrary to eliminativism, *moderate ontic structuralism* maintains that both objects and the relations between them exist at the fundamental level. This version aims to retain the same motivations as eliminativism without the objections raised against it (see also Chapter 6). In contrast to a more traditional view, where objects can exist independently of the structure they are related by, in moderate structuralism, objects are only characterized by the relations they stand in, and these relational properties constitute their essence. While in eliminativism, relations are fundamental and objects are emergent, here both relations and objects are regarded as fundamental. Initially, this view suggested that objects do not possess intrinsic properties. However, in response to criticisms, it was later amended to include them (Esfeld 2004, Esfeld and Lam 2011).

A similar view, which does not include intrinsic properties, is known as *relational holism*. According to this approach, objects instantiate relations rather than intrinsic properties, and these relations are irreducible—they do not depend on any intrinsic features of the individuals that constitute those systems. Thus, it is not really a variety of holism as traditionally defined (i.e., a view in which the whole is more fundamental than its parts) (Teller 1989).

Another alternative to structuralism is *wave function realism,* which will be discussed in Chapter 3. This view is based on the idea that standard quantum theory is too vague and imprecise to safely extract a metaphysics from it. In contrast to structuralists, wave function realists emphasize the dynamics: they aim to solve the measurement problem by proposing a unique dynamics that applies to both the microscopic and macroscopic levels. Furthermore, they argue that structuralists are too empiricist, and they seek to endorse a robust form of realism that describes phenomena at all levels, not just on the macroscopic observable scale. However, as I argue in the next chapter, their approach also turns out to be radical and revisionary in both ontology and explanation. One might think that this indicates a quantum revolution is unavoidable. Nevertheless, I will argue in Chapter 4 that this is not the case: there is a perfectly "nonrevolutionary" way to understand quantum theory as a constructive theory that preserves symmetries.

Further Readings and Discussion

For more on the formalism of quantum mechanics, you can refer to the SEP entry (Ismael 2021), as well as several chapters in Bricmont (2016), Norsen (2017), Maudlin (2019a), and Barrett (2020), to name a few. Feynman's lecture notes (1963) are also an excellent resource. Volume III, on quantum mechanics, can be found at https://www.feynmanlectures.caltech.edu/III_toc.html. A thorough overview of quantum logic is provided in Wilce (2021). For decoherence, see Bacciagaluppi (2020), and for the uncertainty principle, consult Hilgevoord and Uffink (2016).

Further information on the philosophical issues raised by quantum theory is available in Myrvold (2022), including discussions on nonlocality. Notable discussions on this topic are also found in Goldstein et al. (2011) and Myrvold et al. (2021), as well as Norsen (2017) and Maudlin (2019a). For the EPR experiment, see Fine and Ryckman (2020), and for a collection of works on Bell's theorem, check out Bell and Gao (2016).

The most important references for structural realism are found in Ladyman (2020), which provides an overview of various structuralist perspectives and their objections. Other notable books on this perspective include Ladyman and Ross (2007) and French (2014). For further reading, see also Frigg and Votsis (2011), McKenzie (2017), as well as Bokulich and Bokulich (2011) and Landry and Rickles (2012).

3

The Measurement Problem and Wave Function Realism

In Chapter 2, I introduced standard quantum theory, its formalism, and one way in which it could be interpreted from a realist point of view: structuralism. The basic idea was to take standard quantum mechanics at face value and then extract from it the least assumptive metaphysics. According to structuralism, the fundamental ontology is structure, embodied in its laws and symmetries. The wave function is understood as part of this structure. Objects, both microscopic and macroscopic, emerge as useful patterns through a top-down, principle-like, approach. Structuralists explicitly recognize this and argue that it is the proper interpretation, as it reflects the humblest realist perspective. Nonetheless, some critics view this approach as too vague and imprecise, as it is based on standard quantum mechanics. Moreover, even if it works, structuralism seems to support a weak type of realism, since the explanation it provides is too similar to instrumentalist approaches. For this reason, *many realists advocate for a more robust approach*, which involves a unique dynamics and a precise foundational language that does not rely on the notion of measurement. This approach requires solving the measurement problem before "extracting" metaphysics from physics.

Chapter Schema Details

In the first section, I delve more deeply into the *measurement problem* and present some of its most promising solutions. I then discuss a prominent view on quantum metaphysics in this context: *wave function realism*. According to this perspective, the wave function represents a physical field, existing in a high-dimensional space, which becomes the fundamental arena of the world. As a result, the three-dimensionality of space and macroscopic objects are not fundamental but suitably emerge from the wave function. In my view, the strongest motivation for adopting this approach is that it is the only way to *preserve both locality and separability in the fundamental space.*

After presenting the positive reasons for endorsing this view, I turn to its *weaknesses*. I argue that a significant challenge for this perspective concerns symmetry properties. There are two issues in this regard. First, symmetries play a crucial role in at least one version of wave function realism, as they are used to derive the three-dimensionality of space. Therefore, one must have a clear definition of what symmetries are and why they hold. The second problem is that, from the wave function realist standpoint, *quantum theory loses symmetries*, such as Galilean or time-reversal invariance. This is problematic because, as discussed in Chapter 1, invariances allow us to decouple phenomena from our descriptions of them. In any case, if symmetries are not important in this approach, it becomes puzzling why one should care about Lorentz invariance, as proponents of this view have done.

Even setting symmetries aside, I discuss a second major challenge for wave function realism related to the type of explanation it provides. Wave function realists initially focus on dynamics: they aim to solve the measurement problem and provide a precise dynamics for both macroscopic and microscopic systems. One would expect wave function realism to offer a dynamical explanation of phenomena along the lines outlined in Chapter 1. Instead, they do not: *their explanation is a peculiar hybrid*. I argue that this brings wave function realists closer to empiricism than they might want, similarly to structuralists, who explicitly sought to avoid such a position. I show that, in terms of the type of realism they endorse, structuralists and wave function realists are more similar than each would like to admit (presumably). Both end up explaining phenomena either entirely in terms of principles (as structuralists do) or through a peculiar explanatory hybrid. This leads me to the next chapter, where I propose a realist framework that does not sacrifice almost anything we are familiar with—three-dimensional objects, compositionality, and dynamical explanation—while still respecting symmetries.

The Schrödinger Cat Problem

Many realists view structuralism as unsatisfactory because it attempts to offer a realist interpretation of standard quantum mechanics, in which measurements are a fundamental part of the formalism. Structuralists treat measurements as emergent patterns, characterized by their usefulness. As we have seen, this leads to a top-down approach to explanation, fitting well with principle theories. Thus, they can explain the regularities in the phenomena, but they cannot explain why these regularities occur. In this sense, they do not endorse a very strong form of realism. A more robust realist would not tolerate vague concepts in the formulation of a theory. They would seek a precise and unified dynamics for all levels of description, in contrast to what is found in structuralism. Robust realists would argue that *a theory containing imprecise language*—such as the term "measurement"—*is not yet ready to be taken metaphysically seriously*. They believe it is premature to extract an ontology from the standard formulation of quantum theory, as many structuralists attempt to do (this does not apply to moderate structuralism, which has been advocated in theories that solve the measurement problem; see also Chapter 6). In other words, standard quantum mechanics is inadequate as it stands and should not be used as a guide to metaphysics. A Schrödinger-evolving wave function generates unobserved macroscopic superpositions, and even though these are technically eliminated by the collapse rule, this rule is based on the notion of measurement, which remains scientifically vague. Thus, the idea is that *we need to develop new, precise, observer-free and measurement-free ways of solving the problem of unobserved macroscopic superpositions*. This issue is commonly referred to as the measurement problem, or the problem of the Schrödinger cat, as anticipated in Chapter 2.

Unobserved Macroscopic Superpositions

Let me review the problem in a bit more detail than before, based on the version proposed by Schrödinger, who used a cat to highlight the absurdity of the situation (Schrödinger 1935a). Imagine *a cat in a box* connected to a vial of poison, which is triggered by the decay of a radioactive nucleus. If the nucleus decays, the vial breaks, releasing the poison and killing the cat. Assume that all the complete information about a given physical system is provided by the wave function, and that it evolves according to the Schrödinger equation. Then, because of the superposition principle, it is possible for the nucleus to be in a superposition of "having decayed" and "not having decayed" at the same time. This superposition is unobserved because it is microscopic, so it can be considered an unfamiliar consequence of the theory. However, this superposition propagates to the macroscopic world: the nucleus superposition becomes entangled with the vial, which will be in a "broken-unbroken" superposition. This then becomes entangled with the poison, which is in a "released-unreleased" state, and finally, after entangling with the cat, it results in the macroscopic superposition of the cat being "dead" and "alive." This superposition, of the cat being both "dead" and "alive," starkly contrasts with observation, leading many to conclude that it is impossible to describe reality using the tools provided by standard quantum mechanics. In other words, *standard quantum mechanics is empirically inadequate*: it is falsified by the evidence that cats are either dead or alive.

This problem of unobserved macroscopic superpositions is also known as the "*measurement problem*" because it can be rephrased in terms of measurement outcomes. For instance, consider a single particle that can be emitted from a source in two opposite directions, say left and right. Assuming that (a) the wave function provides a complete description, and (b) the wave function evolves according to the Schrödinger equation, it follows that (c) measurements to detect where the particle has gone (left or right) do not yield definite results. If ψ_{left} corresponds to the particle going to the left, and ψ_{right} to the particle going to the right, then their superposition, $\psi_{left} + \psi_{right}$, is also a possible wave function for the particle due to the linearity of wave function evolution. However, this microscopic superposition of the particle propagates into a macroscopic superposition of the detection screen being hit on the left or right. That is, the screen, which is supposed to measure the particle's location, does not function as intended, since the measurement of the particle's position yields no definite result. As discussed in the previous chapter, von Neumann's collapse rule addresses this issue in the standard formulation, but it does so by introducing a new evolution equation for the wave function, which applies when a measurement is performed. But if someone wants a theory that does not involve the notion of measurement in its formulation, what alternatives are available?

Some have proposed that the collapse of the wave function happens in *the mind of the observer* (Wigner 1961). According to this view, only conscious beings can collapse the wave function by interacting with it. This provides a precise theory, but at the cost of depending on a currently unavailable theory of consciousness. Starting in the 1950s, several new quantum theories were developed that did not rely on this idea, offering *different solutions to the problem without involving measurements or observers.* These theories encountered significant resistance within the physics community for a long time (for more on this, see Chapter 7), but today most philosophers of physics recognize their legitimacy. Here, I will discuss only the solutions that seem most promising (for others, see "Further Readings"). All these solutions share the common feature that the dynamics is universal (applicable to everything) and precise, meaning that there are no vague notions in the formalism (such as the appeal to "measurements" or "observers").

The Pilot-Wave Theory

As traditionally presented (Maudlin 1995), *each universal and precise solution to the measurement problem denies one of the following three premises*: (1) the wave function provides the complete description of every physical system; (2) the wave function evolves according to the Schrödinger equation; and (3) measurements have results. *The theory that denies the first premise*—that the wave function provides the complete description of the system—*is known as the pilot-wave theory*. This idea was initially introduced by de Broglie and later refined by Bohm (de Broglie 1924, Bohm 1952). Consequently, it is also referred to as de Broglie-Bohm theory (Valentini 1992, Holland 1993) or Bohmian mechanics (Dürr et al. 1992). In the pilot-wave theory, the complete state of any physical system is described by both *the particles' configuration and the wave function*. The particles move according to a guidance equation, while the wave function evolves according to Schrödinger equation. Though the theory is sometimes presented using the concept of a "quantum potential" that affects particle acceleration (Bohm 1952), it has been argued that it is more accurate to express the theory in terms of an equation that generates the particles' trajectories based on their velocities, defined in relation to the wave function: $v = v(t,\psi)$ (Dürr et al. 1992).

The pilot-wave theory offers a solution to the measurement problem because, although the wave function may be in a superposition state, *the particles always have a definite position*. For example, in the Schrödinger cat scenario, the particles are either in the dead cat's support or the alive cat's support, and thus their behavior is governed by only one part of the wave function. This theory is fully deterministic, both in the equations for the particles and the wave function. The *randomness* observed in quantum phenomena arises from the in-principle impossibility of knowing the exact initial conditions of the particles. I will explore more features of this theory in Chapter 4. For now, it's important to note that, by definition, the pilot-wave theory is *experimentally equivalent* to standard quantum theory when the predictions of the latter are unambiguous. (Standard quantum theory uses operators to represent observables, but in some cases, there is no corresponding operator, forcing standard quantum theory to rely on approximations. In contrast, the pilot-wave theory does not require such approximations. For instance, the measurement of arrival times for particles directed at a target is an example where standard quantum theory struggles due to the lack of a time operator. Standard quantum theory must measure these times indirectly, using different methods that lead to different distributions of possible results. However, in pilot-wave theory, as we will see in Chapter 4, operators are just useful mathematical tools for describing statistics and are not strictly necessary for determining measurement outcomes. As a result, standard quantum theory and pilot-wave theory may yield different empirical outcomes, making them potentially distinguishable. This distinction, however, is highly controversial. For example, there have been ongoing discussions between Goldstein, Tumulka, and Zanghì (2024) and Aristarhov and Das (2024) regarding this matter.)

The Spontaneous Localization Theory

The second approach to solving the measurement problem is to *deny the assumption that the wave function evolves according to a linear equation.* This was, in essence, the goal of the collapse rule, though it was implemented unsatisfactorily. However, it is possible to precisely implement this rule by allowing the wave function to evolve according to an equation that causes it to "spontaneously" collapse in a specific, narrow spatial region, in a manner that is instantaneous and random. In other words, the idea is to *incorporate the collapse of the wave function into the law itself,* thereby eliminating macroscopic superpositions. This is the approach taken by *spontaneous collapse or spontaneous localization theories*, commonly known as GRW theory (Ghirardi et al. 1986) or GRW-P theory, named after the researchers who first proposed and independently contributed to its development (Pearle 1989).

In GRW theory, the wave function evolves according to *a stochastic (not deterministic) and nonlinear dynamics.* The evolution process can be described as follows: the wave function evolves according to Schrödinger equation until a random time is reached, after which it "spontaneously" and instantaneously localizes in a small, random three-dimensional region. The rate of collapse and the width of localization are new constants of nature, which depend on the number of "particles" in the system. These constants are chosen so that *macroscopic objects localize very quickly.* The measurement problem is therefore resolved, as the cat in the Schrödinger thought experiment can remain in superposition, but only for an extremely brief period.

Unlike pilot-wave theory, the GRW theory is intrinsically indeterministic. While nonlinearity alone would be sufficient to suppress macroscopic superpositions, the inclusion of stochasticity is argued to be necessary to avoid superluminal signaling (Gisin 1984). Since the wave function no longer evolves according to the Schrödinger equation, *GRW theory is not empirically equivalent* to standard quantum theory. However, many of the empirical discrepancies between the two are beyond our current ability to detect. Nonetheless, some experiments have been conducted, and so far, they have confirmed the predictions of quantum mechanics. It is worth noticing that the parameters in the GRW theory can be (and have been) adjusted to avoid falsification. Experimental results offer a window of viability for these theories in the space of parameters, but this window is narrowing over time (see Bassi 2007, 2021, and references therein).

The Many-Worlds Theory

A third option for precisely solving the measurement problem is to *reject the idea that measurements must yield definite results*. In this view, one embraces quantum superpositions as fully real at all scales. Each term in a superposition does not merely represent a possible way the world might be; instead, each term corresponds to an actual way the world is. According to this picture, there is no single world, but rather a *multitude of emergent "worlds*," each corresponding to one of the components of the superposition.

These "worlds" *do not interfere* with each other at later times, which prevents phenomena like a dead cat returning to life through quantum interference. This separation is ensured by *decoherence*, as introduced in Chapter 2: when a system interacts with its environment, the coherence between components of the superposition is destroyed, eliminating interference effects. This framework is known as *the many-worlds theory*, or Everettian quantum mechanics, named after Hugh Everett, who proposed the idea in 1957—though he did not articulate it in these exact terms. In the many-worlds framework, the measurement problem effectively disappears because all possible outcomes of a measurement are realized—each in a different world (see Wallace 2012, Vaidman 2021). There is no need for a collapse mechanism or a special role for observation.

It is worth emphasizing that, although these approaches are often referred to as "interpretations" of quantum mechanics, they are, in fact, distinct physical theories. The pilot-wave theory and the GRW theory, for instance, have different mathematical formalisms and provide fundamentally different pictures of reality. This distinction is crucial for the realist who finds structuralism unsatisfying: rather than drawing metaphysical conclusions from standard quantum mechanics—which relies on the vague concept of measurement—realists are urged to take these alternative, more precise theories seriously. These frameworks do not require us to determine whether or not a measurement is being made in order to specify how the wave function evolves; instead, they offer clear, dynamical accounts of physical processes from the ground up.

Wave Function Realism

Now that each of these theories offers a unique and universal dynamics applicable at all levels of description—and that their formalisms rely solely on precise, well-defined notions—quantum theories begin to resemble earlier successful frameworks, such as classical mechanics. From a realist perspective, this marks a turning point: *we are finally in a position to ask what kind of story these theories tell us about the nature of reality*. One prominent answer to this question is *wave function realism*, a view that emerges naturally in this context. According to this perspective, the wave function represents a physical, material field. However, this field is quite unlike classical fields such as the electromagnetic field, which oscillate in three-dimensional space. As discussed in the following, the wave function lives in a high-dimensional configuration space, because it is defined over the configurations of all the "particles" in a system. That is, the wave function is not a wave in ordinary space, but a wave in configuration space. Crucially, in this approach, particles are not considered fundamental. At the most basic level, there is only the wave function. Therefore, it becomes plausible to treat configuration space not as a representation of particle positions, but simply as a label space that encodes the fundamental degrees of freedom of the wave function.

As we will see, there are *alternative approaches* to quantum metaphysics in which the wave function is still treated as representing something material, but not in the form of a field in a high-dimensional space. Perhaps a more precise label for wave function realism would be *high-dimensional space fundamentalism*. Still, the term "wave function realism" has become standard for the view that takes the wave function literally—as describing a real, physical entity—especially in contrast to views that assign it a more instrumental or abstract role. So, despite potential ambiguities, I don't think the use of this label causes genuine confusion in practice.

Preliminary Arguments

According to wave function realism, *the wave function*—understood as a field defined over configuration space—is (or constitutes part of) *the fundamental ontology* of quantum theories (Albert 1996). This is especially evident in the case of theories like GRW and Everettian quantum mechanics, and partially so in the pilot-wave theory. One reason to adopt this view lies in the central role the wave function plays across these theories. *The Schrödinger equation*—or, in the case of GRW, its stochastic, nonlinear modification—*governs the evolution of the wave function*. Just as Newtonian mechanics is about particles because it provides a dynamical equation describing their evolution, so too quantum mechanics appears to be about the wave function, since its central equation governs the dynamics of that object.

A second motivation for wave function realism comes from *quantum entanglement*—long-range correlations between systems separated in ordinary space—which is naturally explained by treating the wave function as a real physical field on configuration space. For example, consider the wave function of two entangled particles, 1 and 2: $\psi_{1,2} = \frac{1}{\sqrt{2}}\left(\phi_1^A\phi_2^C + \phi_1^B\phi_2^D\right)$. This expression describes a superposition in which particle 1 is localized in region A and particle 2 in region C, together with a state where particle 1 is in B and particle 2 is in D. In shorthand, $\phi_1^X \phi_2^Y$ represents particle 1 being in region X and particle 2 in region Y, both in three-dimensional space. Under wave function realism, $\psi_{1,2}$ is interpreted as a field on six-dimensional configuration space (since each particle contributes three spatial dimensions), and it exhibits two distinct "bumps" or concentrations of amplitude. One bump corresponds to the configuration where particle 1 is in A and particle 2 in C; the other corresponds to the configuration where particle 1 is in B and particle 2 in D. These bumps are well-separated in configuration space. In entangled systems like this, by definition *the wave function cannot be decomposed* into a product of separate three-dimensional wave functions for each particle. But if we treat the wave function as a single, unified field in configuration space, then these nonlocal correlations are simply features of the overall structure of the field—features which are directly visible in the configuration space picture.

High-Dimensional Space

Let us see in more detail what the main idea of this view entails. We have said that the wave function is treated as a concrete physical field—similar in some respects to classical fields like the electromagnetic field. For instance, an electric field is a vector field: a mathematical function that assigns a vector to each point in three-dimensional space. Instead, the wave function is not a field over ordinary space; rather, it is *a complex scalar field on configuration space*—a function that assigns a complex number (or equivalently, two real numbers) to each point in configuration space.

So, there are at least two differences with electromagnetic fields. First, a scalar field assigns a single value to each point in a space—like temperature, which assigns a numerical value to each location indicating how hot or cold it is. In contrast, a complex scalar field assigns *two values*: usually interpreted as the real and imaginary parts of a complex number, or alternatively as amplitude and phase.

In addition, and crucially, the wave function is not defined on ordinary three-dimensional space, but on *configuration space*: the space of all possible configurations of the system's particles. For a system of N particles, each with three spatial coordinates (x, y, z), configuration space has $3N$ dimensions. For example, a system with two particles, p_1 and p_2, with positions (x_1, y_1, z_1) and (x_2, y_2, z_2), corresponds to a single point in six-dimensional configuration space with coordinates $(x_1, y_1, z_1, x_2, y_2, z_2)$. In many physical systems, the number of particles can be extremely large. For instance, one mole of a substance contains on the order of 10^{23} particles (Avogadro's number), making configuration space almost unimaginably vast. If the wave function is the fundamental physical entity and it lives on configuration space, then configuration space—not three-dimensional space—is *the fundamental arena of physical reality*, despite appearances to the contrary.

As previously noted, when wave function realists refer to "particles," this language is metaphorical. *At the fundamental level, there are no particles*—only the wave function. What we perceive as particles are *emergent structures* or phenomena derived from the underlying wave function. Similarly, configuration space should not be literally understood as the space of particle positions, but more abstractly as the space of the system's degrees of freedom.

Alternatives

Wave function realism is a *radical departure* from the traditional view that reality is composed of three-dimensional objects interacting within three-dimensional space—a view deeply rooted in classical physics. Some thinkers are untroubled by this shift and even embrace more extreme perspectives. For instance, *Hilbert space fundamentalism* proposes that the wave function is a vector in an abstract Hilbert space. Also according to this view, familiar experiences of three-dimensional space and macroscopic objects emerge from features such as the spectrum of the Hamiltonian (Carroll 2022).

However, some philosophers and physicists view the high-dimensional nature of configuration space as a significant issue. While they might not deny the ontological role of the wave function, they resist interpreting it as a field in high-dimensional space. Instead, they propose alternative metaphysical frameworks that preserve the fundamentality of three-dimensional space. *Structuralist* approaches fall into this category, treating the wave function as part of a fundamental relational structure rather than a high-dimensional entity. Some of these structuralist views, like *eliminativism*, are revisionary in different ways—they deny the fundamentality of objects altogether, treating them as emergent phenomena. Other variations, such as *moderate ontic structural realism* or *relational holism*, hold that objects and relations are equally fundamental. Importantly, no version of structuralism denies the primacy of three-dimensional space in the way that wave function realism does.

A different perspective that retains the fundamentality of three-dimensional space challenges another deeply held assumption: that parts are more fundamental than wholes. Motivated by quantum entanglement again, this view—known as *holism*—suggests that the entire system is more fundamental than its individual components. While structuralists interpret entanglement as evidence for the primacy of relations, holists argue that the wave function should be understood as a property of the whole system. A particularly extreme form of this idea, *priority monism*, asserts that the universe as a whole is ontologically prior to its parts. In this view, parts still exist, but only in a derivative sense (Schaffer 2010, Ismael and Schaffer 2020). A more radical proposal by Howard (1985) suggests that only the whole exists, with nothing else being fundamental. Wave function realism can be seen as a radical form of holism: not only is the whole more fundamental than the parts, but it also resides in high-dimensional configuration space rather than in three-dimensional space.

Other approaches, motivated by a desire to retain the fundamentality of three-dimensional space (or spacetime), have been developed. One such proposal is *spacetime state realism*, formulated within the Everettian framework. In this view, the wave function characterizes abstract features of spacetime regions through their reduced density matrices (Wallace and Timpson 2010).

Another idea, often associated with the pilot-wave theory, is the *multi-field* or poly-wave approach. This treats the wave function as a field in ordinary three-dimensional space that assigns a multi-component amplitude to each point—analogous to how electromagnetic fields assign vectors to points in space (Forrest 1988, Belot 2012, Hubert and Romano 2018). In spacetime state realism, objects are still derived from the wave function, consistent with the Everettian view. In contrast, the multi-field approach maintains that objects are composed of particles, with the poly-wave acting as a kind of quantum analog to classical fields.

Another alternative is *density matrix realism*, which suggests that the fundamental quantum state is a (universal) density matrix, instead of the wave function. This approach has been explored in various contexts, including the pilot-wave theory (Anandan and Aharonov 1999, Dürr et al. 2005a, Maroney 2005, Allori et al. 2014, Chen 2018).

Further concerns about the physical plausibility of a high-dimensional field have inspired attempts to replace the wave function with fields located in ordinary three-dimensional space. One such approach has been developed within the pilot-wave framework and seeks to describe reality *exclusively in terms of local beables*—entities situated in three-dimensional space. However, this strategy encounters a significant obstacle: it requires an infinite number of such fields to recover the full content of the wave function (Norsen 2010a), making the theory highly unwieldy and arguably less natural.

A different perspective is offered by *primitivism*, which maintains that the wave function is simply a fundamental element of the theory—an objective representation of the quantum state that cannot be further analyzed or reduced to more basic components (Maudlin 2019a). In this view, the wave function is "what it is," without needing to correspond to a familiar physical entity or be embedded in a more intuitive ontological framework.

My favorite proposal is what I have labeled *spatiotemporal foundational ontology* approach, even if it is commonly known as the primitive ontology approach. As I will defend in Chapter 4, it holds that, across all fundamental physical theories—including quantum theory—feature a spatiotemporal ontology which serves as the microscopic basis for constructive explanations and respects the symmetries of the theory. Within this framework, the wave function is best understood not as a direct representation of matter or a physical entity in high-dimensional space, but rather as a *functional component* of the laws of nature. As I will argue in Chapter 5, the wave function should be interpreted as a tool that governs the dynamics of the fundamental ontology rather than as part of the ontology itself.

Locality and Separability in the Fundamental Space

Wave function realists are well aware of the alternative interpretations of quantum theory, yet they maintain that, despite its radical metaphysical commitment that fundamental reality exists in a high-dimensional configuration space, their view preserves important theoretical virtues (Ney 2021). Chief among these is the claim that *wave function realism uniquely upholds two intuitive features* that a physical theory ought to possess in its fundamental space: *separability and locality.* As briefly introduced in Chapter 2, separability refers to the idea that a whole is determined by its parts—a kind of vertical or compositional determination. Locality, by contrast, concerns horizontal or interactional determination: it is the principle that what occurs in one region of space does not instantaneously influence what occurs in a space-like separated region. In essence, separability is about the metaphysics of matter—whether objects can be described in terms of their components—while locality pertains to interaction—whether systems can be isolated and whether influences between them propagate at finite speeds.

Wave function realism achieves *separability in configuration space* because the wave function is fully specified by its amplitude and phase at each point in that space. It also satisfies *locality in configuration space* because, within this fundamental space, quantum dynamics do not involve instantaneous action at a distance. According to wave function realists, configuration space is the *true* space in which these features matter, as it is the space in which the fundamental ontology lives. Moreover, they argue that all the puzzling nonlocal correlations we observe in three-dimensional space (discussed in Chapter 2) are rendered straightforward and intelligible when seen as the local behavior of a field in high-dimensional space. In this way, wave function realism offers *a demystification of nonlocality* in physical space by embedding it within a local and separable high-dimensional framework.

Critically, no other major ontological interpretation of quantum mechanics maintains *both* separability and locality in its fundamental space. In fact, apart from my favorite approach, all other views are nonseparable. This is because they treat the wave function as fundamental, and since separability is a thesis about the composition of matter, such views cannot recover the idea that wholes are constituted by parts. My approach is separable because it does not regard the wave function as representing material substance. Within this perspective, compositional reasoning becomes natural, provided the ontology is suitably microscopic. However, my approach is nonlocal: even though we retain separability, the wave function still plays a role in governing the interactions between objects. If two objects are entangled *via* the wave function, then what happens to one can instantaneously affect the other, regardless of spatial separation (again, more on this in Chapter 4). Therefore, wave function realism stands as the only interpretation that preserves both locality and separability in the fundamental space—an achievement its proponents claim should weigh heavily in its favor.

This raises a deeper question: *Why should we care about locality and separability in the fundamental space*? The answer, as Ney (2021) suggests, lies in their *intuitive appeal*. High-dimensional locality enables us to make coherent sense of physical causation and interaction, while separability in configuration space supports Humean supervenience: the idea that all facts about the world supervene on the local distribution of properties. Preserving Humean supervenience is theoretically desirable because it avoids the need to posit extra relational facts, making for a simpler and more elegant metaphysical picture (see later for more). Next, we will critically evaluate whether these reasons—simplicity, explanatory clarity, and preservation of locality and separability—are strong enough to warrant a commitment to wave function realism.

Recovering the World within Wave Function Realism

As previously emphasized, the *counterintuitive aspect* of wave function realism stems from its claim that the world we experience—three-dimensional and composed of familiar macroscopic objects—arises from a fundamental ontology situated in a much higher-dimensional space (Monton 2002, Allori 2013a, b). The difficulty of reconciling this idea with our everyday perception of space and objects is referred to as the "*configuration space problem*." This includes the *three-dimensionality problem*—how three-dimensional space emerges—and *the macro-object problem*, namely how macroscopic objects can be found in the wave function. To address these challenges, wave function realists suggest that three-dimensional space and objects are not fundamental but rather *emergent* or *derivative*. There are at least two major attempts to realize this.

The first approach argues that three-dimensionality emerges from the features of the *Hamiltonian*. Since the "true" Hamiltonian of the world—as represented in the Schrödinger equation—is formulated in terms of three spatial dimensions, three-dimensionality is thereby privileged. Once this three-dimensional structure is identified, three-dimensional microscopic objects can be defined *functionally*. That is, one first defines what it means to be a three-dimensional object in terms of its functional role, and then demonstrates that the wave function can fulfill this role. This functional reduction leads to the emergence of microscopic three-dimensional objects, which can then be *compositionally aggregated* into macroscopic entities (Albert 2015).

However, in my view, this functionalist approach has been successfully criticized by Ney (2021), and therefore I will not pursue it further. Instead, I focus on her account, which uses *symmetries*—particularly permutation invariance—to derive three-dimensional space. Ney employs the notion of partial instantiation to identify what we conventionally refer to as particles within the wave function. But let us start with the symmetry argument. Suppose we have two wave functions, ψ_1 and ψ_2, respectively, localized around the points P_1 and P_2 in the high-dimensional configuration space. For $N = 12$, let $P_1 = (-1,1,0,1,1,0,-1,-1,0,1,-1,0)$; $P_2 = (1,1,0,-1,1,0,-1,-1,0,1,-1,0)$. Now, assume a three-dimensional representation of these points: $P_1 = (-1,1,0;1,1,0;-1,-1,0;1,-1,0)$; $P_2 = (1,1,0;-1,1,0;-1,-1,0;1,-1,0)$. In this framework, four three-dimensional "particles" appear to emerge for each point. By contrast, using a two-dimensional representation: $P_1 = (-1,1;0,1;1,0;-1,-1;0,1;-1,0)$; $P_2 = (1,1;0,-1;1,0;-1,-1;0,1;-1,0)$, we get six two-dimensional "particles." Now consider *permuting* the "particles"—i.e., swapping their positions. Permutation invariance dictates that the behavior of the system remains unchanged under such swaps. This principle relates to the notion of indistinguishability discussed in Chapter 2: since experimental results are what ultimately matter, and these are expressed via the square of the wave function's modulus, swapping particles does not affect outcomes: $|\psi(x_1,x_2)|^2 = |\psi(x_2,x_1)|^2$. In other words, permutation invariance is a symmetry of the dynamics. Crucially, when one *reshuffles* the "particles," *the state remains unchanged—except for their labels—only in the three-dimensional representation.* Thus, only the three-dimensional decomposition preserves the permutation invariance intrinsic to the high-dimensional dynamics. This leads to the conclusion that three-dimensional space is privileged and therefore can be said to emerge nonfundamentally.

From this foundation, we can derive three-dimensional microscopic particles as derivative parts of the wave function. If we speak of particles, it is merely for convenience; such language does not reflect a commitment about what fundamentally exists. What we call *"particles" are lumps in the squared amplitude of the wave function*. These do not necessarily have definite positions, as the wave function may be spread out. Instead, *"particles" partially instantiate different locations to varying degrees*, depending on the squared amplitude of the wave function at those points. This indeterminacy of position allows for a straightforward explanation of quantum interference. In fact, particle-like behavior arises only when the wave function is sufficiently localized. When it is not, wave behavior dominates, such as in interference phenomena. Macroscopic objects, composed of many such emergent particles, maintain stable positions due to semi-classical dynamics—a result of the governing Hamiltonian. In GRW theory, collapse dynamics ensures that superpositions effectively vanish, localizing the wave function in three-dimensional regions. In Everettian mechanics, decoherence—the interaction between system and environment—prevents terms in the superposition from interfering (see Figure 3.1).

This project of extracting particles from the wave function is, in certain respects, very similar to what Schrödinger aimed to do. As we saw in Chapter 2, he treated particles as superpositions of different wavelengths forming a wave packet. This constitutes a precise dynamical mechanism that supports an *effective particle ontology*: at a given scale, it appears *as if* matter is made of particles, provided the wave packets remain localized for a sufficient amount of time. In this sense, wave packets can be thought of as combining like Lego bricks to form macroscopic entities, as long as they do not spread out—namely, as long as they behave like particles. This stands in contrast to what occurs in wave function realism: here, one must first recover three-dimensional space, and only then effectively recover particles, using metaphysical notions such as functionalism or partial instantiation. *These are convenient principles rather than dynamical mechanisms, as in the case of Schrödinger.*

In any case, *Schrödinger abandoned his initial idea* of thinking of particles as wave packets (1928) for several reasons. First, as we have seen, he realized that his wave function was defined in configuration space. He attempted to resolve this by proposing that matter is represented by a three-dimensional charge density field, obtained by suitably integrating the squared amplitude of the wave function. However,

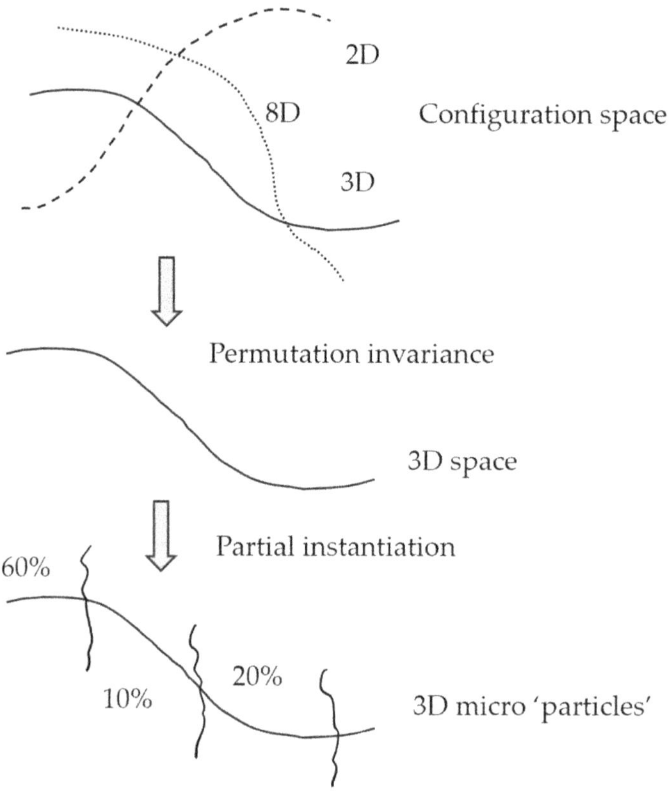

Figure 3.1 Emergence in wave function realism.

he noticed that the wave packet did not remain localized for long, so the appearance of trajectories would not persist. Additionally, due to the superposition principle, different parts of the wave function—even if initially separated—would eventually interfere with one another, leading to a return of wave-like behavior when one expects particle-like behavior. That is, we have the measurement problem: linear evolution of leads to unobserved macroscopic superpositions, which is physically untenable. As we know, one way to resolve this is by *adopting a many-worlds* interpretation: all possible states occur but are mutually invisible because they no longer interact (due to decoherence). Schrödinger presumably saw this result as a *reductio* of the theory, which may explain why he abandoned it. In this light, wave function realism can be seen as an attempt to make sense of Schrödinger's early vision of a wave ontology. However, wave function realists explicitly reject a dynamical strategy for extracting particles from the wave function, thereby relinquishing the aim of providing a constructive explanation (see the end of the next section).

Objections against Wave Function Realism

Wave function realism, in contrast to structuralism, begins with theories that aim to solve the measurement problem. This is because it is maintained that only by eliminating vague notions—such as that of measurement—from the formulation of the theory can quantum theory be made compatible with realism. Proposals like *structuralism or the Information-Theoretic (IT) approach*, which take standard quantum theory at face value without addressing the measurement problem, are seen as *too anti-realist*. Nonetheless, I will argue that, ultimately, wave function realists find themselves closer to the structuralists than presumably they would like to admit.

Before doing so, however, let me discuss *some general challenges* that wave function realism is thought to face. To anticipate, some objections appear more pressing than others. For instance, critics have questioned whether wave function realism can be empirically confirmed or whether it can be extended to a relativistic framework. I believe that these concerns can be satisfactorily addressed by wave function realists (as has been done by Ney 2021), provided they can successfully recover three-dimensionality from configuration space—a point that has itself been contested. However, I believe a more serious challenge lies in the fact that, within this framework, quantum theory appears to lose important symmetry properties. Additionally, the explanatory strategy adopted by wave function realism is a peculiar hybrid of constructive and principle-based explanation, and it remains unclear whether its advantages truly outweigh its drawbacks.

Empirical Incoherence

Assuming one can successfully recover three-dimensionality from the high-dimensional fundamental space (a point I will challenge later in this section), a primary objection is that wave function realism *undermines the very possibility of empirical confirmation*. In other words, even if the approach is true, it would not be capable of being supported by empirical evidence. The concern is as follows: the evidence for a theory is encoded in measurement results, which manifest as pointers indicating specific locations on a graduated scale. However, if wave function realism is correct, such evidence is not given in terms of three-dimensional pointer locations, since pointers themselves are constituted by the wave function and the wave function is not situated in three-dimensional space. Therefore, it is concluded that there would be no possible evidence to confirm quantum theory if wave function realism were correct (Maudlin 2007b).

However, as we discussed in the previous chapter, the wave function realist has a ready response. *Measurement results*—like macroscopic objects and everything else we perceive—do *exist, but they are not fundamental*. They are features of the wave function. If the wave function realist succeeds in deriving macroscopic objects from the wave function, they can also succeed in deriving measurement apparatuses that appear to point in three-dimensional space, thereby defusing this challenge.

The Tension with Relativity

Other challenges for wave function realism are connected to relativistically invariant theories. Before continuing, one might observe that relativistic quantum theories, such as quantum field theories, have yet to solve the measurement problem; thus, arguably, they are not satisfactory physical theories for wave function realists to consider. Nonetheless, one might still want to examine whether wave function realism can be compatible with current relativistic quantum theories, even if these theories are likely to be modified in the future. To properly evaluate the objection, it seems essential to first understand what it means for a theory to be relativistic. Setting aside whether one might want more than this (see Chapter 4), a relativistic theory is one whose laws are Lorentz invariant (see also Chapter 1).

One might think that *wave function realism is in tension with relativity because it is formulated in configuration space*, while relativity is fundamentally a theory of spacetime (Wallace and Timpson 2010). Nonetheless, if wave function realists are successful in deriving spacetime from configuration space, this objection disappears: what is needed is that spacetime exists—not that it exists at the fundamental level.

Otherwise, it has been argued that, since in quantum field theories *the wave function can be mathematically derived in terms of spatiotemporal regions*, the wave function cannot be more fundamental than spacetime, as wave function realism maintains (Myrvold 2015). However, as wave function realists promptly respond, mathematical dependence does not imply *ontological priority*.

Another issue arises from the fact that achieving relativistic invariance entails a *loss of "narratability"*—there is no unique account of how the wave function evolves in time; rather, its evolution depends on the frame from which the phenomenon is observed (Albert 2015). It has been argued that this narratability failure is surprising only within wave function realism, which is local and separable, whereas in other approaches, it is a consequence of nonseparability or nonlocality (Wallace and Timpson 2010). In reply, wave function realists have claimed that even if narratability failure is evidence for spacetime nonseparability, such nonseparability is *still compatible* with high-dimensional separability: it is possible that, while the wave function is separable in high dimensions, it is nonseparable in lower dimensions. Therefore, they argue, the objection evaporates. Nonetheless, I think the wave function realist's reply here is too quick. While it is true that low-dimensional nonseparability might emerge from fundamental high-dimensional separability, this fact alone carries little weight. As we have seen previously, evidence underdetermines theory, meaning that, in principle, one can adjust anything to make everything compatible with everything else—as emphasized by Quine and Duhem. Multiple theories can accommodate the same data: some entail narratability failure, while others do not. However, this does not mean that all such explanations are equally satisfactory. In other words, narratability failure is to be expected in nonlocal, nonseparable theories, but it is surprising from the perspective of wave function realism. According to certain accounts of explanation—such as the deductive-nomological model and its variants—a phenomenon is explained when it becomes expected, when it can be predicted. If so, wave function realism arguably does not explain narratability failure, whereas other interpretations do.

It has also been objected that wave function realism is *not well-defined* within relativistic extensions of quantum mechanics, such as quantum field theories. The problem lies in the difficulty of interpreting current relativistic quantum field theories as theories of a field in configuration space. For instance, one objection concerns the fact that in quantum field theories, *the number of particles is not fixed*—they can be created and destroyed. Since wave function realism interprets the wave function as a field in configuration space, and configuration space has $3N$ dimensions (where N is the number of particles), configuration space would then have a varying number of dimensions. Additionally, in relativistic quantum field theories, systems cannot be represented in terms of a Schrödinger-evolving wave function, making wave function realism unsuitable for such theories. Moreover, it has been argued that privileging position, as wave function realists do, may conflict with the fact that in quantum field theories, observables are often defined in the momentum basis (Wallace and Timpson 2010, Wallace 2020). Ney argues that these objections stem from a *misunderstanding* of the proposals. The point of wave function realism is not to interpret the wave function as a field in configuration space in *all* circumstances. Rather, *the aim is to find a local and separable ontology for a given theory.* In nonrelativistic quantum mechanics, this is what the wave function in configuration space provides, but no one insists that the same ontology must be preserved in a different theoretical context. In other words, a local and separable ontology need not necessarily take the form of a field in configuration space written in the position representation. Having said that, I think it should be noted that moving in this direction places wave function realism in tension with the *principle of metaphysical continuity* (Myrvold 2015, 2019). This principle holds that our metaphysical conclusions should remain robust under small perturbations to the theory itself. If a metaphysical view is highly sensitive to slight modifications in the underlying theory, we lack justification for accepting it as valid. One should, therefore, be cautious in metaphysical speculation, endorsing only those conclusions that remain stable across theoretical changes. This avoids overcommitting to current theories and reduces the risk of requiring constant metaphysical revisions as physics progresses. Adhering to this principle also promotes a stable and resilient metaphysical framework, ensuring that metaphysical commitments do not drastically shift. It further supports unification across theoretical models and fosters a coherent overall picture of the world. However, wave function realism appears *too fragile under theoretical change*—especially if Ney is right in claiming that the wave function is not necessarily the correct ontology for quantum field theories.

The Irrelevancy of High-Dimensional Locality and Separability

More importantly, the motivation for wave function realism—namely, separability and locality—seems questionable to me. As we have previously seen, the strongest argument for wave function realism is that it is the only alternative that is both local and separable. It is argued that we should value these features because they are intuitive. While I am convinced that we should favor intuitive theories (due to their convenience, fruitfulness, explanatory power, and so on), *I am not entirely persuaded by the argument that high-dimensional locality and separability are intuitive*, and therefore worth prioritizing.

First, the locality in wave function realism is locality in the space of the wave function—i.e., configuration space—so quantum theory remains nonlocal in three-dimensional space. The wave function realist emphasizes that locality is needed to make sense of physical action, arguing that nonlocality is counterintuitive: since causes precede effects, we generally understand interactions between distant objects as occurring via influences that travel at finite speeds. That is, we conceive of interactions as local. Nonetheless, this reasoning supports three-dimensional locality, since our observations are of three-dimensional interactions. Therefore, this argument cannot be used to support wave function realism. Moreover, people have traditionally assumed locality—again, in three dimensions—because it allows us to think of objects as effectively isolated. However, the relevant notion here is three-dimensional locality, not high-dimensional locality, since in high-dimensional space, nothing is truly isolated.

Similarly, wave function realists value separability because it preserves our intuitive notion of compositionality: if separability holds, composite systems can be broken down into simpler parts, and the features of the whole can be understood as the sum of the features of its parts. Yet once again, it seems that the separability we intuitively care about is three-dimensional. That is what is familiar to us—we model and manipulate systems under the assumption that the world is separable in three-dimensional space, just as we assume locality in three dimensions. Historically, people have valued separability because thinking of macroscopic objects as composed of microscopic constituents has been extremely successful. But again, this applies to three-dimensional separability, not necessarily to its high-dimensional counterpart. Moreover, it seems puzzling to be so concerned with preserving compositionality if one does not ultimately use compositional explanation. Ney is aware of these difficulties and justifies the desire for locality and separability on the grounds that *they help preserve Humean supervenience,* which is simple enough to be worth preserving. However, this defense is unlikely to convince non-Humeans, and presumably, even many Humeans would prefer a more principled justification.

Circular Three-Dimensional Emergence

Another pressing objection is related to the strategies used to recover three-dimensionality from symmetries (and this also applies to the functionalist account developed by Albert). The concern can be stated succinctly: if one uses symmetries to privilege three-dimensionality over other possible dimensional representations, then it must be clear what symmetries are, why they hold, and why they matter. I will argue that the answers to these questions presuppose three-dimensionality, making the account *circular*.

To understand this better, let us discuss symmetries. Very generally, as explained in Chapter 1, symmetries are transformations that leave something unchanged or invariant. Symmetries are properties of a theory in the sense that the theory's laws remain invariant: a transformation is called a symmetry if it does not alter the description of the phenomena. If this is the case, the theory is said to be invariant under that transformation. More precisely, *a theory is invariant under a symmetry when the symmetry transforms solutions of the theory into other solutions*. For example, a solution $r(t)$ might represent a ball falling in Chicago, Illinois. Consider T, a possible symmetry transformation—say, a translation. Then the theory is invariant under T if and only if $T\left[r(t)\right]$ is also a solution. That is, $T\left[r(t)\right]$, the solution transformed under T, represents another valid physical scenario, such as a ball falling in Cairo, Egypt. To say that the theory is invariant under translation means that the behavior of the ball falling in Cairo is governed by the same laws as the ball falling in Chicago. Symmetries thus express the idea that the description of the same phenomenon from different reference frames (linked by a specific symmetry transformation) should remain unchanged.

Now, recall how the wave function realist uses symmetries to recover the three-dimensionality of space as a derivative concept. The idea is that *permutation invariance should hold*: if we swap the labels of two particles, the law remains unchanged. It is then argued that this symmetry is *preserved only in the three-dimensional representation*. That is, the invariance in the high-dimensional dynamics corresponds to a symmetry that manifests only in three dimensions—thereby singling out three-dimensionality as special. But let us now ask a critical question: *Why is the law permutation invariant?* The answer remains obscure. Indeed, I believe the opposite of what the wave function realist claims: the symmetry in three-dimensional space is the one we observe, and it is because of this observed symmetry that we impose invariance on the laws in high-dimensional space. In other words, *the reason we postulate invariance in the high-dimensional theory is that we want to preserve the symmetry observed in three-dimensional space*—not, as the wave function realist asserts, that the three-dimensional symmetry emerges because of invariance in high-dimensional space. Put differently, we impose high-dimensional invariance because we are trying to safeguard three-dimensional symmetry. This is because, in constructing a theory, we demand that it possess symmetries in three dimensions, aiming to achieve a perspective-free (or frame-independent) description. Therefore, the appeal to high-dimensional symmetries as a foundation for recovering three-dimensionality appears to *beg the question*—it already assumes the primacy of the very three-dimensional perspective it seeks to derive.

Lack of Symmetry Properties

Now I wish to argue that even if one were to successfully justify high-dimensional invariance in a noncircular way—so that symmetries genuinely succeed in privileging the three-dimensional representation—there would still be an inconsistency in attributing importance to symmetries. In fact, on the one hand, this account places paramount importance on symmetries for recovering three-dimensionality. On the other hand, however, if the wave function is a configuration space field, then the theory turns out to be *violating many symmetries*, such as Galilean invariance or time-reversal invariance. In fact, in order to preserve a symmetry, the wave function should transform as to be a possible solution. Nonetheless, there is no reason to think it would transform in the way needed to preserve these symmetries.

Let us see this in more detail. As discussed in Chapter 1, there is a symmetry in classical physics called the pure Galilean transformation, or Galilean boost, which consists in a *translation by uniform motion*. It maps the description of a given system into another that differs from it only by being in constant relative motion with respect to the first. Mathematically, this entails the substitution of $r \to r - vt$. For instance, the description of a falling body by someone in a lab is transformed by a pure Galilean transformation into the description of the behavior of the same body as seen by someone on a train moving at constant speed relative to the lab. Classical mechanics is Galilei invariant: the descriptions of the falling body in the lab and on the constantly moving train are the same. As anticipated, for a variety of reasons, *symmetries are seen as theoretical virtues*—that is, features that, when possessed by a theory, make it superior to alternatives, as discussed in Chapter 1. I have argued that, among other reasons, symmetries are desirable because they allow us to keep the description of phenomena as perspective-free as possible. Regardless, it is a fact that symmetries are often used as guides in the discovery of new theories or in selecting among competing ones: symmetries are treated as requirements and imposed on general models, with those lacking them being discarded. By this reasoning, it would be desirable for (nonrelativistic) quantum theory to be Galilei invariant. However, if wave function realism is true, it turns out that quantum theory is not Galilei invariant. Consider, for simplicity, a single-particle system evolving freely from the point of view of the wave function realist (the case generalizes without difficulty to multiple particles). The wave function is a scalar field ψ in configuration space, which in this case is simply three-dimensional space (since there is just one particle). *How would such a field transform under a pure Galilean transformation*? Think of temperature, which is an example of a scalar field—a function assigning a number to every point in space, $T = T(r)$. If you board a train and move away with constant velocity from a region described by $T(r)$, you will not describe that region as changing temperature merely due to your motion. That is, after the Galilean transformation $r \to r - vt$, the temperature field still satisfies $T(r) \to T(r - vt)$. This is how a scalar field naturally transforms under a pure Galilean transformation: *its value remains unchanged*. For this reason, the wave function in the framework of wave function realism, being a scalar field, should also remain unchanged under a Galilean transformation: $\psi(r)=\psi(r-vt)$. However, mathematically, if $\psi(r)$ is a solution of the Schrödinger equation (i.e., a possible way the world can be), then $\psi(r-vt)$ is not. In order for the transformed ψ to be a valid solution, it must transform as follows: $\psi(r)\rightarrow e^{\frac{i}{\hbar}\left[mvr-\frac{1}{2}mv^2t\right]}\psi(r-vt,t)$. That is, an additional exponential factor must multiply the transformed wave function. In other words, *for quantum theory to be Galilei invariant, the wave function must transform in a way that is incompatible with its nature as a scalar field.* Only when the wave function transforms with this exponential factor do solutions remain solutions. Yet, there is no physical justification for this additional factor: it would be as if temperature changed depending on whether we are uniformly moving with respect to it or not. Therefore, quantum theory is not Galilei invariant, if wave function realism is true (Allori 2018b).

As another example, consider the *time-reversal transformation*, in which t changes sign. As discussed in Chapter 1, classical mechanics is time-reversal invariant (which is precisely what generates tension with the thermodynamic arrow of time), and so it seems desirable to preserve this symmetry in quantum theory as well. To do so, a time-reversed solution must also be a solution. A solution to the Schrödinger equation is a function of time, $\psi=\psi(t)$, which can be pictorially imagined as a set of instantaneous snapshots taken at different times and "glued" together, like a movie. A solution is this movie—a possible history of the world. Each snapshot contains what exists in the world at that instant: the instantaneous fundamental ontology. In this case, being a scalar field, the wave function should simply shift in time without changing in amplitude: $\psi(t)\rightarrow\psi(-t)$. This makes intuitive sense—there seems to be no reason to think that the instantaneous ontology changes depending on whether we view the snapshot from a forward or backward projection of the movie. That would be like saying that the 1979 movie *Star Trek: The Motion Picture* always features William Shatner, whether played forwards or backwards. Nonetheless, $\psi(-t)$ is not a solution to the Schrödinger equation: the "backward movie" with Shatner is not a possible description of the world. Rather, the wave function must be transformed into its complex conjugate: $\psi(t)\rightarrow\psi^*(-t)$. But that would be like saying Mark Hamill plays the backward version of the movie, while Shatner appears only in the forward version—an outcome that would be totally appalling (see Figure 3.2, with Shatner being a star and Hamill a lightning bolt).

One possible way out is to notice that the root of the problem lies in the assumption that the wave function, understood as a scalar field, is part of the material instantaneous ontology—that is, what William Shatner and Mark Hamill are made of. Instead, if

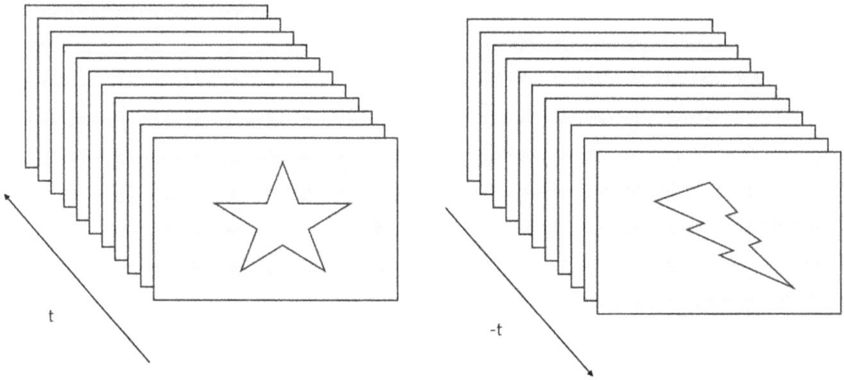

Figure 3.2 Time reversal in wave function realism.

something else represents matter, we would not need to worry about how the wave function transforms. We could then allow any transformation that ensures the actor remains the same in both the forward and backward movie. In this way, the theory could be time-reversal invariant, and the snapshots would behave sensibly (Allori 2019a). (Callender 2020 and Struyve 2022 have argued that one could make sense of time-reversal symmetry within wave function realism; however, as argued in Allori 2025, this logic does not extend to other symmetries.)

To summarize, if we think of the wave function as a scalar field representing matter, then it must transform as such—but as a consequence, quantum theory loses (at least) Galilean and time-reversal symmetries. This is already problematic, since symmetries have historically played an important role in theory discovery and evaluation. As we've seen, they are also necessary to guarantee a description of phenomena that is as independent as possible from our perspective. Of course, one could *bite the bullet* and accept this consequence. However, this seems problematic, especially in the context of wave function realism, where symmetries are crucial for recovering three-dimensional objects. Regardless, notice that if one accepts the loss of symmetries and argues they are not worth preserving, then it seems natural to also give up on the search for Lorentz-invariant theories. If so, however, why are physicists still searching for them? What would it even mean for a theory to be *relativistic* in this context, if not Lorentz invariant?

To conclude, observe that all alternatives to wave function realism that treat the wave function as a physical field—whether in configuration space or not—face the same problem with symmetries. This is because none of them offers a valid reason why the wave function should transform in the way needed to preserve symmetries while simultaneously representing the ontology of matter. This problem does not arise for my favorite approach, discussed in the next chapter, because in that framework, the wave function is not taken to represent anything material. Consequently, there are no constraints on the possible transformation of the wave function under a certain symmetry linked to its nature (a field or what have you). Rather, the only important element is to guarantee that symmetries are preserved, and this is accomplished by letting the wave function to transform accordingly (see Chapter 4).

Hybrid Explanation

We saw in the previous chapter that structuralism can be understood in terms of principles rather than through constructive explanation. This is because, within structuralism, objects—whether microscopic or macroscopic—are considered merely useful patterns. Macroscopic objects are not thought to be composed of microscopic entities, nor are their properties explained in terms of microscopic dynamics. Rather, structure constrains and systematizes the phenomena. In a sense, something similar occurs here: *within wave function realism, phenomena are not explained constructively.* Since the wave function is a field in configuration space, one must first derive three-dimensionality, then microscopic particles, and finally macroscopic objects. The first two steps involve some kind of principle. It is only when one "recovers" microscopic three-dimensional particles that one can finally use a constructive explanation to account for macroscopic objects and their properties, as we already saw. *What are these principles?* In Ney's account, for instance, three-dimensionality is obtained by appealing to a principle based on symmetries: "take as privileged the space which preserves the symmetries of the dynamics." To derive particles, another principle is used, this time based on the notion of partial instantiation: "take as particles those parts of the wave function which behave as particles." In Albert's proposal, where the Hamiltonian is said to privilege three-dimensionality, the situation is even more intricate, as the dynamics enters at the first stage, but still as a principle: "take as privileged the space in which you write the Hamiltonian." Macroscopic objects are then understood functionally, again via a principle: "an object is what an object does."

Evaluation

Structuralists and wave function realists are both realists: they believe that our best theories can tell us about the nature of the world. However, while wave function realists claim that one must solve the measurement problem to begin the realist enterprise, structuralists disagree. They argue that we can interpret standard quantum theory just as effectively, as we saw in Chapter 2. Some structuralists interpret the solutions to the measurement problem in terms of structure (moderate structuralists), but the majority do not. This is because structuralists, like proponents of the IT approach, maintain a *strong empiricist* attitude and are skeptical of the unobservable. Therefore, they are not concerned with grounding their fundamental ontology at the macroscopic level. Let's call this type of realism *weak realism.* It reflects the stance of approaches like the IT interpretation, rainforest realism, or even Qbism: these theories solve the measurement problem by eliminating macroscopic superpositions, using the von Neumann rule to systematize appearances. As noted earlier, these approaches use a *principle-based* explanation: there are certain principles that delineate what can and cannot happen. Although most proponents of these views claim that their approach is realist, many other realists would argue that this position does not meet the criteria of a true realist stance.

Wave function realists seek more than weak realists because they aim for a precise solution to the measurement problem. This is because they wish to *eliminate macroscopic superpositions precisely*, without the vagueness that sometimes emerges in the theory's definition. They accomplish this either through spontaneous, law-induced collapse (as in GRW theory), decoherence (as in many-worlds theory), or by supplementing the wave function (as in pilot-wave theory). "Measurement" is not precise enough a concept: What qualifies as a measurement, if not a physical process involving the interaction of microscopic objects with macroscopic ones?

Now, let's explore *why having precise notions is important*. The typical response is that precision leads to a *unified dynamics*. In contrast to standard quantum theory, where Schrödinger evolution breaks down during measurement, the precise solutions to the measurement problem maintain the same evolution equation at all descriptive levels. We should care about having the same dynamics for all objects because it avoids privileging any particular level of description over others. To achieve this, one needs to push the privileged description to the *microscopic level* and account for the other descriptions in terms of it. This suggests that one should still seek *constructive theories*. However, wave function realists ultimately propose theories that are not entirely constructive. As we have seen, they use principles that systematize the data. In other words, it seems that, contrary to their initial goal, *wave function realists veer toward weak realism*: their approach reproduces appearances by systematizing phenomena. They initially claim they want more: a unified understanding of fundamental reality, akin to that of constructive theories. If they end up doing what weak realists do—systematizing data, but precisely rather than vaguely—let us ask the question again: Why should one care about this precision? Arguably, as just seen, wave function realists care about the dynamics because, unlike weak realists, they are focused on obtaining a precise dynamical equation for all objects at all scales—not using the von Neumann collapse for macroscopic objects and linear Schrödinger dynamics for microscopic ones. Wave function realists also use dynamics (specifically, decoherence) to show that macroscopic objects persist. However, *the way they rely on dynamics is not constructive:* for example, they use the symmetries of the Hamiltonian to recover three-dimensional appearances, whereas a constructive theorist would use the solutions of the Hamiltonian, as these represent a potential way the world could be. Nonconstructive explanations that emphasize dynamics seem to pull in opposite directions: the former pushes toward weak realism, while the latter leads to a more robust form of realism, making wave function realism a peculiar hybrid. There is tension between the wave function realist's desire for a robust form of realism and the kind of explanation their view actually provides, which is not fully constructive. *The wave function realist begins as a robust realist but ends up closer to weak realism.*

If that's the case, one might wonder *what the point of solving the measurement problem actually is* if one can do without it, as structuralists do. The point of solving the measurement problem, as wave function realists claim, is to have a precise rule to specify wave function collapse. But why would one want a precise rule if they are only concerned with appearances? Presumably, they want a unique dynamics. But why is the wave function realist interested in dynamics if they provide a nondynamical explanation (Allori 2023a)?

Moreover, the strongest argument for wave function realism is that it preserves *locality and separability. But why* should we care about these *if the explanation is nonconstructive*? After all, as discussed earlier, separability is the basis of compositionality, and locality is crucial if we care about the dynamics.

One could defend wave function realism by noting that, while it is true the principles are necessary to recover three-dimensionality and microscopic objects, after this recovery, macroscopic objects can be effectively thought of as composed of microscopic entities. In this sense, *constructive explanation is still used*: it is not the only step needed to recover the phenomena, but one of the two—principles first, constructive explanation second. Having at least one constructive step is important. For instance, consider thermodynamics and statistical mechanics. Many argue that the laws of thermodynamics can be constructively accounted for through Newtonian mechanics via statistical mechanics. If a wave function realist lacks a constructive step in their explanation of phenomena, it becomes difficult to see how they could accept statistical mechanics as just outlined. In other words, if someone desires quantum theories to be explained solely through principles, they should not be too attached to a constructive understanding, not just in quantum theory but across the board. Yet, statistical mechanics does constructively explain thermodynamics, and arguably this extends to quantum statistical mechanics. But if wave function realism offers a nonconstructive explanation, there is tension. *How do these two explanations align*? While wave function realists claim to follow Schrödinger's path by taking wave ontology seriously, they do not follow the explanatory mechanism he envisioned, as discussed in Chapter 2. If they did, they would arrive at theories with a many-worlds character, which conflicts with constructive explanation from the outset—likely not what Schrödinger intended.

Finally, like structuralism, *intrinsic properties do not exist* within wave function realism. This could be seen as an objection to the view if one believes that properties are necessary for explanation. Traditionally, fundamental entities are thought to possess fundamental properties, like mass and charge, which allow a given law to act on these objects. These properties define what an object is: an electron, for example, is a particle with a mass of $9.1093837 \times 10^{-31} kg$ and a charge of $1.60217663 \times 10^{-19} C$. These properties also appear in the law and explain how the law affects electrons, as opposed to protons. To explain the behavior of an object, we refer to its properties. This is not the case in wave function realism, where the wave function is the fundamental entity. What are its fundamental properties? Presumably, its amplitude and phase—however, these are properties of the whole, not the parts. We don't use amplitude or phase in the same way we use mass and charge to explain the behavior of a physical system. In Chapter 6, I explore a speculative approach where I eliminate fundamental properties, which could align well with wave function realism.

To conclude, in the last two chapters, I have discussed two ways to approach the realist question within quantum mechanics (what does it mean to make quantum mechanics compatible with scientific realism?). On one hand, we have structuralism, where structure is considered more fundamental than objects, which emerge as useful fictions. On the other, we have wave function realism, which holds that matter fundamentally exists in a high-dimensional space, with three-dimensionality, particles, and macroscopic objects emerging through various techniques. Both structuralists and wave function realists highlight features they consider worth preserving, but their chosen features differ. *Structuralists* argue that their ontology is the most parsimonious, the simplest way to understand and expand modern physics, and the one requiring the least in terms of unobservable entities. These principles—*simplicity, parsimony, fruitfulness, and humility*—are regarded as desiderata, as discussed in Chapter 2, because they are *intuitive*. *Wave function realists* argue that their view is the only one that preserves *locality and separability* in the fundamental space, which are seen as valuable because they, too, are *intuitive*. In other words, each approach argues that a theory is preferable if it possesses certain worthwhile features. For structuralists, these features are parsimony, simplicity, humility, and fruitfulness; for wave function realists, they are locality and separability. Given this disagreement, one must argue either that their chosen features are more important or that their view has fewer costs. As we've seen, structuralism's costs include being highly revisionary, with no fundamental objects, nonseparability, and the failure of constructive explanation. Wave function realism is also revisionary, as it denies the existence of a fundamental three-dimensional space and fundamental objects. Its explanation of phenomena is neither entirely constructive nor based purely on principles. Additionally, it is not humble in the structuralist sense, as it assumes entities beyond measurement outcomes and structures. Structuralists prefer humility over separability, while wave function realists prefer the opposite. Nonetheless, *both approaches fail at providing a fully constructive explanation*, making them explanatory revisionary (see also Allori 2022b). The questions I will explore in the next chapter are: *Can we have a fully constructive explanation of quantum phenomena? What costs are involved? Are they worth it?*

Setting their differences aside, both approaches claim that quantum theory requires a significant revision of our understanding and explanations. While most would describe these views as revolutionary, I argue that they propose *unnecessarily radical metaphysics*. One needs good reasons to adopt such radical views. All things being equal, it seems we should favor the least revisionary strategy. In the next chapter, I will argue that my approach—the spatiotemporal foundational ontology approach (formerly known as the primitive ontology approach)—provides just that: a fully constructive approach. It also has the advantage of positing three-dimensional space and a fundamental three-dimensional ontology. However, it is not humble, as it assumes the existence of microscopic unobservable entities and uses super-empirical virtues to choose among many possibilities. It is also not parsimonious, as it adds other entities to quantum theory. But its lack of parsimony and its "daring" metaphysics are necessary costs to obtain a constructive explanation. This robust realism does not shy away from positing unobservable entities if it leads to a better explanation. More importantly, it is nonlocal, which is its higher (perceived) cost, as discussed in the next chapter. My view is partly motivated by the challenges faced by the approaches discussed in the previous chapters. In Chapter 6, I explore what the wave function is in this framework and highlight its structuralist elements.

Further Readings and Discussion

Material on the measurement problem and the various responses to it can be found in the following books and papers, among others: Albert (1992), Lewis (2015, 2016), Maudlin (2019a), Barrett (2020), Dürr and Lazarovici (2020), Myrvold (2022), Adlam (2021), and Tumulka (2022). Greenberg et al. (2009) provide a comprehensive compendium on quantum theory and its interpretations, while Freire et al. (2022) offer insightful historical perspectives.

In addition to the sources cited in the text, the pilot-wave theory is discussed in Valentini (1992), Holland (1993), Cushing et al. (1996), Dürr and Teufel (2009), Bricmont (2016), and Norsen (2017). Other important review articles include Allori and Zanghì (2004), Tumulka (2021), and Goldstein (2021).

A review of spontaneous collapse theories can be found in Ghirardi and Bassi (2003, 2020), Bassi (2007), Lewis (2005, 2006, 2022), and Allori (2022c). Examples of recent developments are given by Gao (2018) and Allori et al. (2020). Precursory work on this topic includes Diósi (1988) and Pearle (1989). The connection between collapse and gravity has been explored in Diósi (1987, 1989) and Penrose (1996, 2014). The latest discussions on empirical testing are presented in Bassi (2021).

For insightful historical work on Everettian quantum mechanics, see DeWitt (1970), DeWitt and Graham (1973), Byrne (2010), and Barrett and Byrne (2012). Barrett (2018) provides a comprehensive review, while an interesting collection can be found in Saunders et al. (2010). Other notable books include Barrett (1999), Wallace (2012), and Wilson (2020). For more accessible yet accurate readings, see Deutsch (1997) and Carroll (2019).

The literature on wave function realism and the status of the wave function is continually evolving. Notable papers advocating for or exploring wave function realism include Albert (1996), Loewer (1996), Ney (2012, 2015, 2017, 2020, 2021), North (2013), and Lewis (2004, 2013). Criticisms of the view can be found in works by Monton (2002, 2006, 2013), Maudlin (2007a, 2019a), Allori (2013a, b, 2018a, 2023a, b), Emery (2017), and Wallace (2020). For interesting papers on the status of the wave function, refer to the collection edited by Albert and Ney (2013). A thorough review of the various positions on the nature of the wave function is provided by Chen (2019). For more on the status and role of the wave function in quantum theory, see Chapter 5.

4

A Spatiotcmporal Ontology for a Symmetry Preserving Constructive Theory

In Chapters 2 and 3, we have seen how both *structuralists and wave function realists* motivate and defend their views. Structuralists argue that their approach is preferable due to virtues such as simplicity, parsimony, fruitfulness, and humility. By contrast, wave function realists have recently advocated for their view on the grounds that it is the only one which is local and separable in the fundamental space. Their main drawback, as I argued in Chapter 3, is that they are *revisionary both in ontology and in the explanations they provide*: while structuralists actively pursue explanation in terms of principles, wave function realists seemingly start off seeking constructive explanations, but ultimately offer a hybrid account. Furthermore, symmetries are lost in wave function realism. And even though structuralism emphasizes their importance, it provides no explanation for why these symmetries hold.

In this chapter, I present and defend my version of the primitive ontology framework, which I have renamed the *spatiotemporal foundational ontology* approach. According to this view, the fundamental ontology of matter—the fundamental constituents of matter—is represented in the theory by some three-dimensional entity evolving in time, or by a four-dimensional spatiotemporal entity. The motivation for this approach is twofold: *to preserve constructive explanation and to respect symmetries*. These aims are guaranteed, respectively, by the fact that what represents the fundamental material ontology is *microscopic and spatiotemporal*, and by this entity being placed at the *foundation* of the theory's architecture—hence the new name. In the quantum domain, these considerations imply that the wave function never represents matter. I argue that this approach is to be preferred over the alternatives because it is *ontologically less radical and offers greater explanatory power*, in that it provides a constructive explanation grounded in symmetries. We have already seen in Chapter 1 why symmetry-preserving constructive explanation is important: it provides an explanation for *why* a certain phenomenon, rather than another, occurs. Explanations provided by principle theoriers, by contrast, merely tell us *what* to expect, without telling us *why* it happens. As discussed previously, preserving symmetries is also important because they ensure that the description of a given phenomenon is as independent as possible from our particular way of describing it. So, the basic idea is this: *if you think that explaining a phenomenon involves providing a constructive account of why it occurs, and if you think that this description should be as perspective-free as possible, then I argue that you should endorse this approach.*

It will then become clear that there is an obvious choice: *the pilot-wave theory*. Although in the literature I may have contributed to suggest that other quantum theories—with a fundamental spatiotemporal ontology—can also provide a constructive explanation that respects symmetries, I will now argue that, ultimately, it is difficult to justify regarding them as serious contenders. Initially, I justified exploring these alternatives due to their fruitfulness; for example, in the case of GRW-type theories, I previously argued that they are worth investigating because of their seemingly greater compatibility with relativity. However, I have recently come to believe that their costs are higher than I initially realized.

Chapter Schema Details

In the first section, I discuss the idea behind my approach: a fundamental physical theory explains a phenomenon when it provides a *constructive explanation respecting symmetries*. I show how the mathematical framework of a theory, understood in this way, displays a *specific architecture*. At the foundation, one finds a *microscopic spatiotemporal* variable that describes the fundamental material ontology, whose spatiotemporality and microscopicality guarantee a constructive explanation. In addition, the candidate for representing the fundamental ontology needs to be *at the foundation* of the theoretical architecture to guarantee symmetry properties. Variables which are not foundational—even if they are in spacetime—they are better understood as *nomological variables*. These variables ultimately relate to the interactions between physical objects: they serve to generate or describe the law governing the evolution of matter. This hierarchical structure is straightforwardly present in classical mechanics, based on Newton's particle ontology.

My approach seeks to *extend the classical explanatory* schema to all physical theories, including quantum mechanics. In the second section, I discuss the pilot-wave theory—the *simplest (empirically adequate) constructive quantum theory that respects symmetry principles*—where the simplest type of ontology (particles) evolves under the simplest type of law (deterministic). I show how, within this framework, familiar quantum features such as effective collapse, uncertainty, the Born rule, operators, contextuality, and nonlocality can be naturally derived.

I then turn to the question of how to constrain the possible laws for various foundational ontologies, appealing to *super-empirical virtues*. As noted, simplicity clearly favors the pilot-wave theory. One might wonder whether exploring alternative solutions to the measurement problem could yield deeper insights. However, I argue that the *many-worlds* interpretation offers no compelling advantage and appears more like a *last-ditch effort* to avoid falsification. Instead, considerations of fruitfulness and symmetry—especially in relation to *compatibility with relativity*—may initially point toward *spontaneous localization theories*. These can indeed be extended relativistically using only relativistic spatiotemporal structures, unlike deterministic alternatives. Yet, contrary to my initial expectations, the *drawbacks* of such theories may ultimately outweigh their benefits.

Important Side Remarks

As already stated, the view defended in this book is my personal take on—and extension of—what is known as the *primitive ontology* approach. The term was introduced by Goldstein, Dürr, and Zanghì (1992), although they did not elaborate on it in writing, except in Allori et al. (2008). Since I am not the one who originally proposed this framework, I do not claim to speak on their behalf (or on behalf of others who currently defend it). So, please take the arguments in this book for what they are: my best defense of an approach that, in my view, captures what fundamental physics should aim to do in order to fully explain phenomena.

Also, a note on terminology. As anticipated in the introduction, I have come to realize that the label "primitive ontology" can obscure what the approach is actually about. To avoid misunderstandings and confusion, in this and the following chapters, I will use, as much as possible, expressions like "the fundamental ontology of matter according to the theory" instead of "the primitive ontology of the theory," as is more common in the literature. Let's call O the fundamental ontology of a theory—that is, what exists in the world according to that theory. Material things are part of O: they are, indeed, M, the *material ontology* (or the ontology of matter). However, there is more to O: other entities, such as laws or properties, are needed to specify how matter moves. Call these N, the nomological ontology (also called the *nomology*, for short). Thus, in this approach, $O=(M;N)$: the ingredients needed to explain phenomena are M, the fundamental ontology of matter, as well as N, the fundamental nomology. The nomology describes the interaction between material entities, and it consists of the set of variables in the theory that help define the law of evolution for the fundamental material ontology. While the fundamental material ontology is at the foundation of the theory, the nomology is not. For more on the nomology, see Chapters 5 and 6.

The Hierarchical Nature of Theories

In the previous two chapters, we have seen two prominent quantum metaphysical views: *structuralism and wave function realism*. They are *revisionary*: no fundamental three-dimensional space, no objects at the fundamental level, and a nonconstructive way of explaining phenomena. One might think this is *inevitable*—that this is what quantum theory forces us to accept. We are witnessing a quantum revolution and need a radically different way of understanding things. Not a revolution like the one advocated by the anti-realists, who believe that our concepts are inadequate to account for the quantum world, or that observers create reality—but a revolution nonetheless. This is not the first revolution in the history of science, and it will not be the last: some are big, like rejecting the idea that the Earth is at the center of the universe, and some are smaller, like coming to believe that matter is made of atoms, and thus mostly void, rather than a continuum of stuff. This, it is argued, is just how scientific progress seems to work: with better theories, we gradually develop a better understanding of the world. However, in the process, we move from more intuitive to less intuitive pictures (in the sense that they are further and further away from what we perceive), and we come to accept what were previously thought to be radical claims about the nature of the world because they are implied by our best theories. Quantum mechanics, they say, just seems to follow this trend.

In this chapter, *I want to challenge this idea. I argue that we are not required to accept any of these radical implications*. In fact, it is possible—through my preferred approach (Dürr et al. 1992, Allori et al. 2008, Allori 2013a, b, 2015b, c, 2018a, b, 2019b)—to continue using the well-tested constructive explanatory schema used in classical mechanics: there are microscopic fundamental entities in (three-dimensional) space that compose macroscopic objects, whose properties are accounted for by their microscopic dynamics. My framework prescribes the structure that any fundamental physical theory, classical or not, should have in order to provide a constructive explanation of phenomena while preserving symmetry properties. By construction, therefore—and in contrast with wave function realism and structuralism—the image of the world provided by quantum theory according to my approach does not radically depart from the classical image. This creates an important advantage over the competitors, since it allows the use of at least some of the *classical constructive techniques*, with appropriate modifications, even in the quantum domain. Given this, I will argue that the classical-to-quantum theory change does not amount, as is often presented, to a scientific revolution in the Kuhnian sense (Allori 2015b). That said, before we move to quantum theory, let me further characterize the spirit of this approach within the domain of classical mechanics.

Theory Architecture and the Nature of Matter

In general, a fundamental physical theory possesses some *mathematical formalism* containing several variables. Some of these represent the *fundamental material ontology*, and they play three main roles. First, they have a *metaphysical role*: they specify what matter is according to the theory. Second, they have an *explanatory role*: they ground a constructive explanation of the phenomena. And finally, they have a *physical role*: they allow the theory to possess important symmetry properties. To better understand this, consider the formalism of classical mechanics discussed in Chapter 1. The theory can be regarded as postulating that matter is fundamentally made of particles represented by three-dimensional Euclidean points. If *the material ontology is an empirically and theoretically informed postulation*—rather than something we read off from the formalism *a posteriori*—then presumably, one should choose the simplest (in some obvious sense) hypothesis first. In fact, there seems to be no reason to do otherwise, and one may always reconsider if this choice does not play out as desired. In addition, the theory specifies *a law for the temporal evolution of matter*. In the case of classical mechanics, the evolution of the particles is prescribed through Newton's equation, $F = ma$: when forces are present, the velocity of an object changes proportionally to its mass. That is, the theory uses other variables, which we usually identify as masses, charges, forces, potentials, and so on—these acquire specific values depending on the physical situation under consideration. For instance, consider a body of $m = 1\, kg$ falling under the action of the gravitational force of the Earth from a height $h = 1\, m$. It will experience a force $F = G\frac{mM_{Earth}}{R^2}$, where R is approximately the radius of the Earth $(R = R_{Earth} + h \cong R_{Earth})$, M_{Earth} is approximately the combined mass of the body and of Earth $(M_{Earth} + m \sim M_{Earth})$, and G is the gravitational constant. With these specifications, one can solve Newton's equation and predict how the body moves.

This simple observation is enough to show that the different variables in a fundamental physical theory are not on the same footing. That is, *theories have a hierarchical structure*. The variables that describe *the fundamental material ontology are at the foundation of the theory*. Then there are many other variables, including those used to write the law of motion for matter. Some of these are defined in terms of the fundamental variables, like the velocity $v \stackrel{\text{def}}{=} \frac{d}{dt}r$ or the momentum $p \stackrel{\text{def}}{=} mv$. Other variables describe the *interaction* with other things, such as the force F. Regardless, these variables are needed to complete the physical description of the phenomena. In this sense, they are fundamental variables. However, they do not represent material objects. Rather, they concern the temporal evolution of systems, their behavior, and their interactions with other systems, rather than the systems themselves.

Notice that these variables are *not foundational*: they do not represent the building blocks of everything else. Instead, they occupy a higher level within the theory's architecture, when compared with the fundamental material ontology. Because of this, they have *a different role* in the theory. To see this, consider for example the different roles of particle position and particle mass in classical mechanics. Position is a defining feature of matter, while mass can be regarded as a higher-level property of matter (see Chapter 6 for more). One may say that a particle with a given position in space has a particular mass, but the reverse does not make much sense: What would it mean to say that a particular mass has a given position? Consider also position and velocity: What would it mean to say that a particular velocity has a given position? In this last case, in particular, velocity is defined in terms of position—as its variation in time—so it is natural to think of velocity as occupying a different "level" than position within the architecture of the theory. If the foundational variables specify the nature of matter, the other variables serve to "dress up" the fundamental entities so that the theory can account for the phenomena.

In other words, the role of these nonfoundational variables is not to describe objects, but to *prescribe what those objects do*. They allow one to define the motion of the fundamental variables, and by virtue of this role, they are not at the foundation of the theory. For instance, as we have seen, in order to predict where a particle will go (i.e., to solve the equation of motion), in addition to the mass, charge, and position of the particle, one must specify the forces F_{grav} and F_{elect}, or equivalently, the potentials. These potentials are not necessarily descriptive or representational variables: they do not describe material entities or their properties. Rather, they are prescriptive, in the sense that they must be specified to determine the motion of the variables representing the fundamental material ontology. Thus, they are *nomological in nature*: they are part of the ingredients of the law of nature that governs—or otherwise describes—the behavior of matter, rather than part of the description of matter itself.

To summarize: roughly speaking, the variables in the fundamental ontology of matter are *descriptive, and foundational*—they tell us *what there is*. In contrast, the other variables are *prescriptive and nomological*—they *tell matter how to move*. While the former represent matter, the latter are connected to laws, properties, or structure, as we will see in the next chapter (see Figure 4.1).

A Microscopic Ontology in Spacetime for Constructive Explanation

As mentioned above, one may invoke considerations of simplicity when choosing which variables represent the fundamental material ontology. Nonetheless, a more important constraint we wish to preserve is that of *constructive explanation*. Particles seem ideal in this regard: a particle theory is well suited to providing a constructive explanation because we can think of macroscopic objects as being made up of microscopic particles. In addition, we may aim to account for their macroscopic properties in terms of microscopic dynamics, and to deduce, under suitable approximations (and at least for most initial conditions), the macroscopic laws from this microscopic behavior. To use a pictorial and metaphorical slogan, *particles are like tiny Lego bricks* that can be assembled in different ways (Figure 4.2, top). After

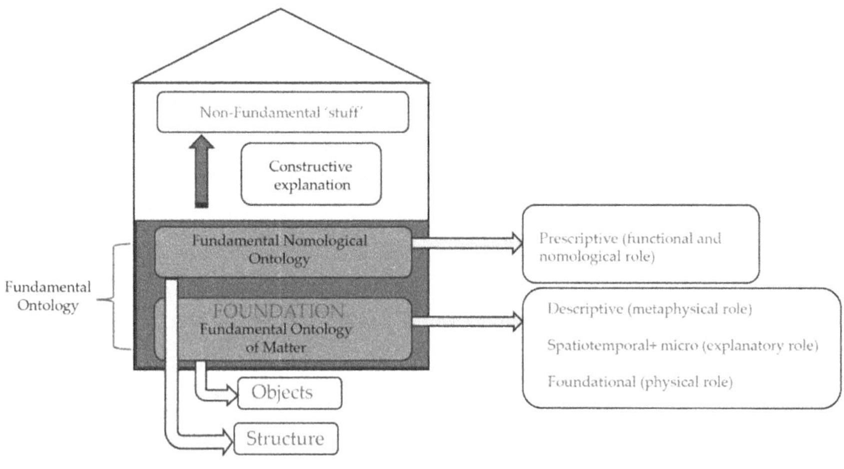

Figure 4.1 Theory architecture.

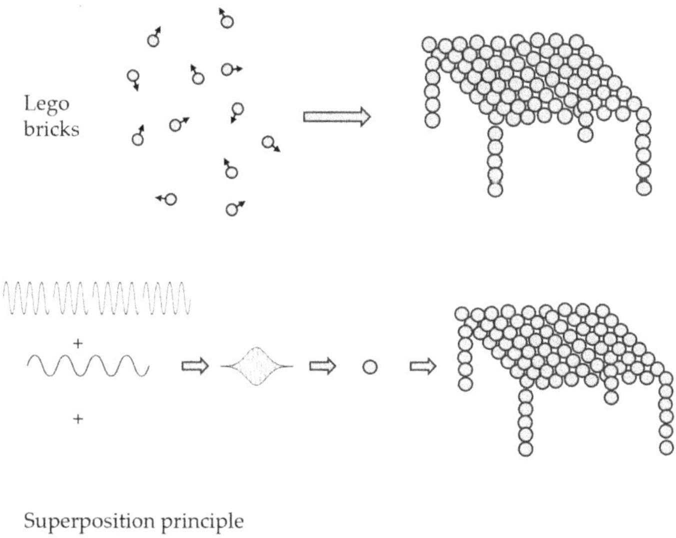

Figure 4.2 Macroscopic objects understood in terms of a particle ontology (up) and a wave ontology (down).

all, this is what happens in classical mechanics: because of its particle ontology, classical mechanics is arguably able to account for every macroscopic property of physical objects using compositional and dynamical techniques. If quantum effects can be neglected, macroscopic solid bodies, gases, and fluids (i.e., tables, water, air) and their properties (the solidity of the table, the transparency of the water, and the temperature of the air) can all be explained in terms of the motion of point-like particles moving in three-dimensional space. That is, the relationship between the image of the world provided by our senses (sometimes called the *manifest image*, Sellars 1967) and the one provided by the theory (the *scientific image*) is *constructively bottom-up*: fundamental entities exist microscopically, and they constitute and determine the macroscopic properties we assign to the complex, macroscopic objects we observe. A theory that explains things in this way is a constructive theory, rather than a principle theory, as discussed in Chapter 1.

There are two features of a particle ontology that ensure a compositional explanation: (1) particles are located in space and evolve in time (i.e., they are *spatiotemporal*) and (2) they are *microscopic*. This stands in contrast to a non-three-dimensional ontology, such as the wave function, which requires more than simply a region of three-dimensional space to be specified. (Some use Bell's 1987 terminology of *local beables*, where "be-able" contrasts with "observ-able," and "local" means three-dimensional. However, since I use the term *locality* to refer to the idea that interaction travels at finite speed, I avoid Bell's terminology to prevent confusion. Moreover, local beables are not necessarily foundational, and thus they do not have the connection to symmetry that is central to my approach—see Allori 2021a.)

Why are these two features—spatiotemporality and microscopicity—so crucial? Because if the *explanans* (the fundamental ontology) is not in the same space as the *explanandum* (the phenomena), then the theory must invoke unexplained principles. In other words, *if the fundamental ontology of matter is not spatiotemporal, then before we can explain the behavior of macroscopic objects, we must first recover three-dimensional space—and this cannot be done constructively but only through principles.* Consider wave function realism: some propose defining three-dimensional space functionally, others via symmetries, but neither yields a fully constructive explanation. Moreover, *if the fundamental ontology is not microscopic, then compositionality fails by default* (though with some qualifications, as discussed below). If what is fundamental exists at the macroscopic level, then by definition, macroscopic objects cannot be understood as composed of more basic entities, nor can their behavior be explained in terms of the dynamics of such components. As we saw with the IT approach or rainforest realism (Chapter 2), explanation in such frameworks proceeds top-down and resembles a principle theory more than a constructive theory.

That said, what we've just discussed applies to ontologies *localized* in three-dimensional space, such as particles. However, particles may not be the only possible spatiotemporal ontology. One could also consider a fundamental ontology that is three-dimensional but *not localized*, such as waves in three-dimensional space. As anticipated in Chapters 1 and 3, such an ontology can still provide a constructive explanation, depending on what physical behavior it is intended to capture. For example, as we saw in the case of light, the fundamental ontology could consist of waves—represented by fields—which are three-dimensional but spread out (i.e., *delocalized*). *What does a constructive explanation look like in the case of a wave ontology*? As anticipated in Chapter 1, the requirement of microscopicity does not make much sense here, because waves are inherently delocalized. Compositionality, too, must be reconceived. Waves do not combine like particles; we cannot think of macroscopic objects as built from microscopic Lego bricks. However, *waves are defined as combining in terms of the superposition principle*, and this is enough to support a constructive explanation: macroscopic phenomena can still be accounted for dynamically in an appropriate limit. This is what occurs in the case of geometric optics, whose equations can be derived from the laws of classical electrodynamics in the eikonal approximation.

The situation changes, as discussed in Chapter 3, when one adopts *a universal wave ontology*, in which wave-like entities must also account for particle-like behavior. In that case, one must show that empirically adequate particle trajectories can be recovered at the appropriate scale as suitably composed from waves. This is precisely what Schrödinger attempted: he proposed that particles could be modeled as *nonspreading wave packets*. In this view, macroscopic objects can be thought of as composed of these wave packets—now functioning as suitably microscopic elements—instead of fundamental particles (see Figure 4.2, bottom).

A Foundational Ontology for Symmetries

So far, we have examined the classical case, in which the fundamental material ontology—particles—is also foundational: it occupies the bottom level of the theoretical architecture. We have discussed why it is important to define the fundamental ontology in spacetime and at the microscopic level—namely, to enable constructive explanation. However, we have yet to clarify the specific role of having a *foundational variable* at the base of the ontology. I previously noted its connection to symmetries, but how exactly does that connection work? One might ask: *What happens if a theory posits a nonfoundational variable as the fundamental ontology of matter*? Exploring this thought experiment will help illuminate the significance of foundational variables within a theory's architecture. In particular, it will clarify why it matters that the fundamental ontology of matter is grounded in such a variable.

Sometimes, a nonfoundational variable can be immediately ruled out by other criteria—for instance, because the variable is not three-dimensional, as in the case of the wave function. However, in other situations, this is less obvious. If a nonfoundational variable *is* three-dimensional, one might wonder whether it could qualify as a candidate for fundamental material ontology. Take, for example, *electromagnetic fields*: these are mathematically defined as functions *on three-dimensional space*, rather than a higher-dimensional configuration space. Because of this, one might naturally think of treating them as *fundamental*. However, we must remember why fields were introduced in the first place: they were posited to account for the motion of particles. *They serve the same role as potentials do*—they help implement the motion of the particles. Therefore, like potentials, fields are best understood as *nonfoundational*, and hence not part of the fundamental material ontology.

Now, I want to argue that there is an additional, related reason to consider fields as part of the nomology rather than the ontology: if we do not, *the theory loses important symmetry properties*. To see this, suppose we *do* include fields as part of the fundamental ontology of matter, and consider what happens under a time reversal transformation $t \to -t$. As discussed in Chapter 3, such a transformation should *not* alter the amplitude of the fields—there's no compelling reason to expect that the magnitude of a field should differ depending on whether it belongs to the forward or backward temporal evolution of a system. Yet, if the amplitude of the fields remains the same, the time-reversed fields are no longer solutions of the theory. In other words, time reversal would cease to be a symmetry of the theory. To preserve the symmetry, we would have to impose a special transformation rule for fields under time reversal: in particular, the magnetic field B would need to flip sign. But once again, we must ask: *Why should this happen*? Recall the analogy from Chapter 3: if something in the ontology changes *merely* because it appears in the forward or backward movie, this is akin to claiming that William Shatner—who plays Captain Kirk in the forward movie—transforms into Mark Hamill playing Kirk in the backward movie. This implies that fundamental objects change their very nature simply by being part of a snapshot from the forward or backward view of the world. That seems completely unwarranted. By contrast, if we treat the fields not as part of the fundamental ontology of matter, but rather as part of the nomological structure of the theory, then they are free to transform however necessary to preserve the theory's symmetry properties. The snapshots—what we take to represent the world's fundamental structure—remain unchanged. Since the fields are not part of what is fundamental, they can adapt their transformation behavior accordingly. In sum, *nonfoundational three-dimensional variables, when mistaken for fundamental material ontology, jeopardize the symmetry of the theory*. Recognizing their proper nomological role is essential for maintaining both a constructive explanatory structure and relevant invariance properties.

Since symmetry properties are crucial—they ensure a perspective-free description—we must impose additional constraints on the possible fundamental ontologies of matter. By analyzing the theory's architecture, we find that *only foundational three-dimensional ontologies can preserve the required symmetries*. Put differently, consider the class of three-dimensional ontologies that are effectively microscopic—such as point particles or any spatiotemporal ontologies that behave, at some intermediate scale, like particles (e.g., wave packets). Among these, only those capable of preserving the theory's symmetries are viable candidates for the fundamental ontology of matter within this framework. Since the fundamental ontology is what the theory ultimately describes, *the symmetries of the theory are determined by the symmetries of its fundamental ontology*. A theory is said to be invariant if the evolution of its ontology is invariant under a given transformation. To ensure such invariance, the higher-level variables—those not part of the foundational level—must transform appropriately, as illustrated by the behavior of electromagnetic fields. These variables adapt in order to preserve the symmetry of the overall system, reinforcing their nomological rather than ontological status.

As previously noted, this position directly contrasts with approaches that treat the wave function as part of the fundamental ontology. These frameworks regard the wave function as ontological, even though it is not foundational, and this results in the loss of key symmetry properties (Allori 2021a). For example, as we already saw in Chapter 3, in addition to failing to preserve time reversal invariance, wave function realism—as well as any view that treats the wave function as representing matter—also violates Galilean invariance. Recall that for a theory to be Galilei invariant, the wave function must transform under a Galilean boost in a peculiar way, rather than as $\psi(r) \to \psi(r - vt)$. However, there is no justification for why a material field would transform in this way: this would be to claim that the description of a physical field depends on whether we are viewing it from a uniformly moving or stationary frame. But that would be unjustified: a material field should be invariant under a change in inertial frame, transforming simply as $\psi(r) \to \psi(r - vt)$. Yet such a transformation does not preserve Galilei invariance. Hence, any framework that treats the wave function as a material field fails to maintain this symmetry.

At this stage, an objection might be raised: *Aren't you relying on symmetry principles in your reasoning, and doesn't this compromise the commitment to constructive explanation?* (Lewis, p.c.). I would argue that it does not. It's essential here to distinguish symmetry principles from the broader category of principle theories. Principle theories attempt to explain phenomena by postulating constraints that restrict what is physically possible—they are meant to account for what happens in the world by invoking general laws or principles. *Symmetry principles*, by contrast, *play a meta-theoretical role*. They do not aim to directly explain physical phenomena. Instead, they serve as heuristic tools for evaluating and choosing between competing theoretical frameworks. In this context, a symmetry principle might be articulated as follows: "Prefer the theory that exhibits a higher degree of symmetry." This is not a claim about what we observe in the world, but rather a criterion for theoretical adequacy. It doesn't describe or predict phenomena, but it does help guide us in constructing explanations that are more consistent, elegant, and objective. Furthermore, this principle is not arbitrary or aesthetic—it carries deep conceptual weight, which provides a strong motivation to use it. The more symmetrical a theory is, the less its description depends on the perspective or frame of reference of the observer. That is, *symmetry enhances objectivity*, offering a more invariant account of physical reality. Thus, far from betraying the goals of constructive explanation, the use of symmetry principles actively supports the pursuit of objectivity in science. They help ensure that our theories reflect the world itself, not merely our vantage point on it.

Comparison with the Alternatives

For the reasons reviewed in Chapter 3, almost everyone agrees that the worldview depicted by quantum theory is radically different from the one emerging from classical physics. Because of this, many have identified the transition from classical to quantum mechanics as a prototypical example of a *paradigm shift*, with the rise of quantum mechanics amounting to a scientific revolution. Initially, Bohr argued that our concepts are hopelessly incomplete. Later, Wigner and von Neumann, concerned by the presence of macroscopic superpositions, proposed that these superpositions are never observed because the observer plays an active role in the theory: it is the act of observation that collapses the wave function into one of the components of the superposition—either by consciousness (as Wigner suggested in 1961) or by the physical act of measurement (in von Neumann's 1932 formulation). Neither approach resembles the paradigm of classical mechanics, thus supporting the view that a paradigm shift has occurred. In both cases, there is no constructive explanation: macroscopic properties are not derived from microscopic processes, but are instead determined either by macroscopic actions (as in von Neumann's view) or by a nonphysical entity (the observer's mind, in Wigner's version). Structuralism also frames the classical-to-quantum transition as a paradigm shift: it portrays reality in terms of structure, not individual objects. Likewise, wave function realism constitutes a paradigm shift, with fundamental space no longer being three-dimensional, matter possessing a holistic character, and the need to develop new techniques to recover classical appearances from the quantum realm.

Nothing of this sort occurs within my framework. On the contrary, under this interpretation the wave function does not represent material objects, and—just like in classical theories—the world is described by microscopic three-dimensional entities, which we can conceive of as composing macroscopic objects. Other interpretations have also emphasized the importance of a three-dimensional ontology: as briefly discussed in Chapter 3, this includes the multi-field approach, priority monism, density matrix realism, existential holism, and the theory of exclusively local beables, including quantum state primitivism. However, none of these alternatives adequately accounts for why quantum theory exhibits the symmetries it does, as demonstrated at the end of the previous section. If one accepts that symmetry is important, this serves as additional support for my approach.

We will explore the status of the wave function in greater depth in Chapter 5. Still, it is worth saying more here about *primitivism* (Maudlin 2013, 2019a). According to this view, the quantum state exists and constitutes the fundamental ontology of matter. The primary motivation behind this position is that the wave function is objective and explanatory, and therefore must be a feature of reality. However, this view *leaves many questions unanswered.* For example: What exactly does this quantum state represent? Should we posit a new type of matter, similar to how in classical electrodynamics we thought of matter and light as distinct—even though, unlike in electrodynamics, the quantum state is not three-dimensional? If so, is the space in which the wave function lives also fundamental? And if this quantum field and matter reside in different spaces, how do they interact? Why is there no reciprocal influence of matter on the wave function? One possible way out of these puzzles is to grant that the wave function is objective without assuming it is also material. It's important to distinguish these two claims: to say that something is objective is not to say that it represents something material. Instead, the wave function might represent a law, or a property, or some nomological element of the theory. Furthermore, simply providing an explanation does not imply materiality—laws offer perfectly valid explanations, yet we do not regard them as material entities. Indeed, electromagnetic fields were long viewed as useful explanatory fictions, and even though the quantum state appears in the laws, this does not make it material—any more than the potentials appearing in laws make them material. As a side note, this line of reasoning suggests that the standard *ontic vs. epistemic* categorization of the wave function may be imprecise or uninformative. The distinction is intended to capture whether the wave function expresses our knowledge of a system (epistemic) or provides an objective description of it (ontic) (Harrigan and Spekkens 2010). However, something may be ontic, in the sense of being mind-independent and objective, without representing something material. On this point, I agree with primitivists that the wave function is ontic, but I disagree with their conclusion that it must therefore be material.

To summarize, a three-dimensional ontology allows us to ground a constructive explanatory framework that—though suitably qualified—is continuous with the classical one. In this sense, there is *no true quantum revolution* (or at least, not the kind that is often advertised). This is especially evident in the case of particle ontology, which provides a strong counter to anti-realist arguments, such as the pessimistic meta-induction. Since particle ontology is preserved through the transition from classical to quantum mechanics, and it underpins the empirical success of both theories, we have good reasons to believe in the reality of particles (Allori 2018a). For more on these points, see Chapter 5. In agreement with Emery (2017), and following the principle that—*ceteris paribus—one should prefer a theory whose scientific image deviates least from the manifest image*, my framework is preferable to its alternatives. It does not require a radical metaphysical shift, and it achieves no less understanding, while preserving much more continuity with both classical physics and common sense.

A Quantum Foundational Framework

As presented in Chapter 2, in contrast with classical theories, it is not clear what the underlying metaphysical assumption in quantum mechanics is. What is matter made of, according to quantum theory? Putting anti-realism aside, as we have seen, structuralism and wave function realism offer their respective answers. According to structuralism, fundamentally there is only structure, and both microscopic and macroscopic objects emerge as real patterns. In contrast, wave function realism holds that the wave function, a non-three-dimensional entity, is fundamental.

If one were *to relax the microscopicality* of the fundamental ontology of matter, perhaps one could argue that the IT approach posits a macroscopic spatiotemporal fundamental ontology. Indeed, measurement outcomes play a special role in this interpretation: the laws make predictions about them and govern their evolution—quantum mechanics is about them. However, if the fundamental ontology of matter is no longer microscopic, then one can no longer maintain that it constitutes the relevant Lego bricks out of which macroscopic objects are built. This would likely necessitate a top-down, *nonconstructive explanatory framework*. Similarly, as in structuralism, if objects themselves are not part of the ontology, then this kind of bottom-up, constructive explanation must be relinquished.

On the other hand, if one sacrifices the three-dimensionality of the fundamental ontology of matter, one might still view the wave function—understood as wave function realists do—as a viable possibility. Yet again, this would undermine the central idea of my approach: the *constructive schema* only works if the fundamental building blocks are objects that inhabit the *same space* as the macroscopic entities whose behavior we aim to explain—that is, three-dimensional space. If the fundamental building blocks of nature exist in a different space—as is the case in wave function realism—then one must take an additional explanatory step to account for how we appear to live in three-dimensional space while, in fact, we do not (as discussed in Chapter 3). This move demands the introduction of principle-based explanations.

Moreover, note that the *principle of metaphysical continuity* introduced in Chapter 3—according to which the best metaphysics is the one that remains resilient under small theoretical changes—is fully compatible with this framework. This principle serves as a safeguard against speculative metaphysical commitments, encouraging us to adopt ontological conclusions that are more likely to endure through scientific development. It fosters a robust metaphysics. In my approach, the fundamental ontology of matter resides in spacetime, as it does in classical mechanics, and is therefore stable, not vulnerable to minor shifts in theoretical formulation.

In short: a material ontology without objects (as in structuralism), a macroscopic spatiotemporal ontology (as in the IT approach), and a non-three-dimensional ontology (as in wave function realism) all require nonconstructive approaches to explain the macroscopic world. My perspective avoids this, whenever possible. *To preserve constructive explanation while also respecting symmetries, one must eliminate the wave function from the ontology of matter and require the fundamental material ontology to be foundational, three-dimensional, and microscopic.* But how, then, can this idea be extended to quantum theory, if the wave function cannot represent matter?

The response becomes clear when we turn to one of the proposed solutions to the measurement problem, namely *the pilot-wave theory*, due to its particle ontology. From my point of view, this is ideal. Indeed, it appears to be the obvious choice for a quantum theory that satisfies the desiderata I am advocating: constructive explanation and preservation of symmetries.

The Pilot-Wave Theory

Given its particle ontology, the situation in the pilot-wave theory is very similar to classical mechanics. One can speak of macroscopic objects as being composed of microscopic particles, and one can explain macroscopic properties in terms of the dynamics of these microscopic constituents. The explanation is, fundamentally, constructive. As Sheldon Goldstein often puts it (p.c.), it is "*OOEOW*: *the Obvious Ontology Evolving in the Obvious Way*." Jean Bricmont likewise characterizes this theory as the rational completion of quantum mechanics, since it simply adds what is needed to standard quantum theory to turn it into a theory of matter moving in spacetime (2016).

What, then, is the wave function in this approach? The next chapter is devoted to addressing this question in detail. For now, let me briefly note that the wave function does not represent the fundamental ontology of matter. Rather, it describes the interaction between particles—it realizes one of the ingredients required to specify how the fundamental ontology of matter evolves. In this sense, *the wave function is a nomological variable*: its role is to help implement the law of motion governing the behavior of the fundamental ontology of matter.

Because of this, the *symmetries* of a theory are determined by the fundamental ontology of matter, not by the wave function. Roughly speaking, to say that a theory has a certain symmetry is to say that the symmetry transforms one possible way the world could be into another such possible way. Accordingly, just as with electromagnetic fields, the wave function is free to transform in whatever way is useful to preserve the symmetry properties of the fundamental ontology. For instance, the pilot-wave theory is Galilei invariant if and only if the particle trajectories—as well as the transformed trajectories—correspond to possible states of affairs. This invariance is achieved by allowing the wave function to transform in the specific way discussed earlier—a transformation that, however, is hard to justify within the framework of wave function realism. In contrast, within my framework, the wave function, being a nomological variable, occupies a higher level in the architecture of the theory. Its purpose is not to serve as a foundation, but to support the implementation of the law of evolution for the fundamental ontology. Since symmetry is a property of the solutions of a theory, the wave function can transform in whatever way is needed to preserve Galilei invariance (see also Allori 2022e). The same reasoning applies to time reversal invariance: for the pilot-wave theory to be invariant under time reversal, the wave function must transform into its complex conjugate—and it does so, precisely in order to preserve the symmetry. As I have already emphasized, preserving symmetries is crucial for enabling a description of phenomena that is as perspective-free as possible (see Chapter 1).

Emerging Quantum Features

The previous section showed that the simplest, most straightforward, constructive quantum theory is the pilot-wave theory. In this section, I briefly review how *all the quantum features* we are familiar with arise naturally when this theory is understood within my proposed framework.

To begin with, the *collapse* of the wave function, one of the postulates of standard quantum theory, emerges in suitable circumstances within the pilot-wave theory. In this theory, the wave function of the universe is given by $\Psi(x_1,\ldots,x_N)$, where $x_1,\ldots,x_N$ are the configurations of all existing particles, and it evolves according to the Schrödinger equation. The wave function of a particular subsystem, composed of particles $x_1,\ldots,x_i$, is obtained by inserting the actual configurations—denoted by capital letters—of the particles not included in the subsystem (i.e., the environment) into the universal wave function. That is, for $Y = X_{i+1},\ldots,X_N$ the conditional wave function is defined as $\psi(x_1,\ldots,x_i) = \Psi(x_1,\ldots,x_i,Y)$. In general, the wave functions of systems smaller than the universe do not evolve according to the Schrödinger equation. However, in suitable situations, they may approximately do so, enabling one to treat different systems as if they were isolated. In a measurement situation, by contrast, the system–apparatus interaction is strong enough that the wave function no longer evolves via the Schrödinger equation. Instead, it effectively behaves *as if* it collapses into one of the terms of the superposition. From that moment onward, the motion of the particles in the system is governed, for all practical purposes, by this "effectively collapsed" wave function. While the specific term into which the wave function collapses is determined, we cannot predict which one it will be, because the result depends on the initial positions of the particles—information to which we have no access. This is a consequence of quantum equilibrium, as discussed by Dürr et al. (1992).

In this respect, the *uncertainty principle* and the *Born rule* can both be understood by analyzing what it means to acquire information, and what limitations quantum theory imposes in this regard. In a deterministic theory, perfect knowledge of every particle's initial conditions and the laws governing them would, in principle, allow perfect prediction. However, acquiring knowledge about a system requires amplifying its features to the macroscopic level—for example, by coupling it to a macroscopic system we can access directly. Such a system is typically called a "measurement apparatus," because we believe it allows us to extract information about the microscopic system. For instance, if we place an object on a properly calibrated scale, the object interacts with the scale, causing a pointer to move and indicate a mass label like "1 *kg*." In this way, we infer that the object weighs 1 *kg*. Obviously, if the system and apparatus are independent, there is no interaction, no correlation, and thus no knowledge of the system. This is crucial to understanding why quantum theory appears probabilistic, even if it might be deterministic at a deeper level. In fact, independence of systems is a hallmark of equilibrium, understood as in statistical mechanics: when nothing changes, no information is exchanged. To gain information, we need nonequilibrium. If the world is governed by pilot-wave theory, then the configurations of particles are in "quantum" equilibrium with respect to the wave function—which itself is not in equilibrium. Just as in thermodynamic equilibrium one cannot extract useful information, in quantum equilibrium the only accessible information about particles is that provided by the wave function. That is, we cannot know more about the positions of particles than what the system's wave function provides. This limitation of knowability is what we usually refer to as the uncertainty principle. The pilot-wave theory also correctly accounts for experimental results, which are statistical in nature: it yields probability distributions of outcomes rather than deterministic predictions. This follows from our ignorance of the initial particle positions. Crucially, assuming an initial universal wave function and an initial random distribution of particle positions given by $|\psi_t(x)|^2$, the distribution of particles will remain of that form under time evolution—this is known as *equivariance*. Therefore, when performing an experiment to determine the position of a particle, the probability $P(x)$ of finding it at position x is given by the square modulus of the wave function evaluated at that point: $P(x)=|\psi(x)|^2$. As an (imperfect) analogy: just as in statistical mechanics a system tends toward greater entropy, here a system tends toward regions where the squared modulus of the wave function is larger (Bricmont 2016).

Furthermore, it can be shown that experimental results can be effectively represented using matrices or *linear operators*, as in matrix mechanics. As in standard quantum mechanics, each experiment can be associated with a suitable operator, $\widehat{A}$, whose eigenvalues correspond to possible experimental results. If the wave function is written as a linear combination of $\widehat{A}$'s eigenstates, then the probability $P(a_i)$ of obtaining a particular result a_i is given by the square modulus of the coefficient c_i corresponding to that eigenstate: $P(a_i)=|c_i|^2$ —this is the generalized Born rule.

It is important to note, however, that *most experiments do not actually reveal anything about the system:* by interacting with the system, they perturb it too much. We might refer to those rare experiments that do not alter the system significantly as genuine measurements, since they can be thought of as revealing a property of the system as it existed prior to the measurement. Experiments that do change the system cannot properly be considered measurements. In classical physics, it is often safe to assume the system–apparatus interaction can be neglected. For example, when measuring temperature with a thermometer, the reading represents an equilibrium between the object and the instrument. Because the thermometer is designed so that this equilibrium temperature is close to the object's initial temperature, we can interpret the reading as revealing a pre-existing property. However, this assumption breaks down when the measurement instrument is comparable in size to the system. For instance, to determine the position of an unmoving particle, we must see it—that is, we must illuminate it with photons, which scatter and disturb the particle. The more accurately we try to determine its position, the more we perturb its velocity, thereby limiting the precision with which it can be known as the uncertainty principle states.

Let us apply this to an actual case. In standard quantum mechanics, spin is often treated as a paradigmatic quantum property. However, spin turns out to be *contextual*: its value depends on how the measurement is performed. That is, the spin value of a particle changes depending on whether it is measured alone or with another property. To see how counterintuitive this is, imagine that measuring the length of a table along with its weight yields a different result than measuring its length with its height. This raises serious concerns if one thinks of spin as an intrinsic property: What kind of fundamental property changes its value like that? These concerns have led many to view spin as another paradoxical aspect of quantum mechanics, inspiring much debate (see, e.g., Mermin 1993). However, within the pilot-wave theory understood in my framework, there is no mystery: spin is not a property. Rather, it is an ingredient in the law governing the motion of matter. In a spin experiment, we are not measuring spin; we are changing the system (see Norsen 2014). The fact that the result changes depending on the experimental setup is simply a consequence of the fact that the apparatus modifies the system—possibly destroying the very property one is trying to observe. Imagine measuring the length of a table by disassembling its legs and comparing the surface with another object. If you then try to measure its height, the result will differ from what you would have obtained beforehand. Whether an experiment is destructive or not depends on the context. In classical systems, it is often possible to perform *nondestructive experiments*, whose results genuinely reflect pre-existing properties. In quantum systems, by contrast, many experiments are destructive and should be interpreted as revealing information about the interaction, not the system itself. As discussed, experimental statistics can be represented by operators. Sequences of destructive experiments are represented by *noncommuting operators*, meaning the order matters: $\widehat{A}\widehat{B} \neq \widehat{B}\widehat{A}$. This makes sense—if experiments are destructive, it matters which is performed first. If you read a message and then burn it, it is different from burning it and then reading it. Understanding spin measurements as destructive interactions between microscopic systems and macroscopic apparatuses removes the mystery of contextuality and clarifies that spin is not a property (Daumer et al. 1996, Dürr et al. 2004, see also Chapter 6).

Let me close with a remark on *the role of hidden variables*. As discussed in Chapter 2, the EPR paper proposed completing quantum mechanics with hidden variables to explain observed perfect correlations locally. For this to work, the hidden variables must be noncontextual (Schrödinger 1935b, Bricmont et al. 2020a, b, 2022)—that is, their values must remain the same across all measurements. However, theorems by von Neumann (1932), Gleason (1957), Bell (1966), and Kochen and Specker (1967) show that noncontextual hidden variable theories that reproduce all quantum predictions are logically impossible. These no-go theorems demonstrate that *hidden variables cannot provide a local explanation of quantum correlations* if they match the predictions of quantum mechanics (which they must, given empirical violations of Bell's inequality). This result falsifies Einstein's statistical interpretation, in which the wave function describes an ensemble, while each individual system is characterized by noncontextual hidden variables (for more, see Allori 2026). In the pilot-wave theory, all properties except particle positions are contextual. But this is not problematic: it simply reflects the fact that at the quantum level, interaction—even when intended as "measurement"—alters the system. So, either properties exist but are destroyed by measurement, or they don't exist at all (except for positions). In the former case, they are useless for explanation; in the latter, what we call "properties" are actually part of the law of motion (see Chapter 6). How does the pilot-wave theory explain the perfect correlations in EPR? Not through hidden variables (which merely specify the spacetime material ontology), but through what Einstein deemed unthinkable: nonlocal interaction—to be discussed in the next section.

Quantum Nonlocality

As seen in Chapter 2, there are nonlocal correlations in quantum theory (not considering superdeterministic or retrocausal theories; see Chapter 7). Separability is about objects, while locality is about the interaction between objects. In standard quantum theory, where the wave function alone describes everything there is, *entangled states are nonseparable*: a composite system in a singlet state has a definite spin property, but its parts do not. The violation of Bell's inequality has shown that *there are nonlocal correlations* between certain arbitrarily distant systems. Each metaphysical view has its own way of accounting for these features, as briefly sketched in Chapter 3. In particular, wave function realists argue that their main advantage is that their view is local and separable—even if in high-dimensional space. Since the objects of our perceptions are fundamentally in the wave function and they interact through it, it is not surprising that both separability and locality are properties of the wave function. This is indeed a straightforward way of explaining why we have nonlocal three-dimensional correlations and lack separable properties: these are the traces of the fundamental high-dimensional ontology. For wave function realists, this is the evidence we actually have of the fundamental ontology being high-dimensional. On the opposite side of the spectrum, we find structuralism, in which, since an ontology exclusively of relations is postulated, separability and locality are properties of the structure. In this sense, one can see structuralism as an example of monism too—structure-monism—since the structure represents both the objects and their interaction. Because of this, also in this case, locality and separability (or the lack thereof) go together: the wave function is nonseparable, in the sense that we may be unable to assign properties to parts, and nonlocal, as the wave function also is responsible for carrying the interaction.

My approach is in the middle: *separability and locality are separate because objects and laws are separate. The fundamental material ontology* lives in a three-dimensional world, like structuralism, but in contrast with it, it *is separable*, because there is more than just structure: macroscopic bodies are composed of microscopic entities. Instead, *the interaction*, carried by the wave function, *is nonlocal*. In other words, within my approach, the straightforward option to account for the nonlocal correlations is to take them at face value: they are about the interaction between objects rather than the objects themselves. While in wave function realism the observed low-dimensional nonlocal correlations and the nonseparability of properties are seen as shadows of the high-dimensional local ontology, and while in structuralism the structure is both nonlocal and nonseparable because structure is three-dimensional, in my view one can employ a *divide et impera* strategy: the fundamental ontology is separable, while the nonlocality remains in the interaction between locally defined entities.

Let me briefly recall that the *world looks local because decoherence*—namely, the interaction with the environment—destroys the quantum correlations. One can show that, in the suitable classical limit, particles are effectively guided by locally plane waves, and classical trajectories emerge as the limit of the pilot-wave trajectories (Allori 2016).

As anticipated, *nonlocality is the most troubling aspect* of this theory from my perspective. The main reason for this concern is that nonlocality is in contrast with relativity, as discussed next. As we saw in Chapter 2, Bell, following EPR, tried to see whether one could provide a different type of understanding of quantum phenomena (including the nonlocal quantum collapse), assuming locality to be true. He proposed a theory in which there is some generic "hidden" ontology, whose dynamics is governed by a Schrödinger-evolving wave function. Experiments ruled out such a theory, in the sense that they showed the locality assumption to be unwarranted.

The Compatibility with Relativity

What does it mean that a theory is relativistically invariant? This question has no uncontroversial answer. As mentioned at the end of Chapter 1, the relevant invariance in relativity is under *Lorentz transformations*, which are transformations involving both space and time, due to the fact that the velocity of light is constant in all (inertial) frames. In my framework, the invariance of a quantum theory under a given transformation concerns the invariance of the law for the wave function only indirectly, contrary to what is often believed. This is because symmetries are symmetries for the law describing how matter moves. So, for relativistic invariance, the law for the fundamental ontology of matter should be invariant under Lorentz transformations. Nonetheless, details aside, people agree that the wave function evolution should also be Lorentz invariant, in order to ensure that the relation between energy and momentum is relativistic. As a consequence of the Lorentz transformations, energy E and momentum p, which classically were connected by the relation $E = \frac{p^2}{2m}$, are now connected by this relation: $E^2 = (pc)^2 + (m_0c^2)^2$, where m_0 is the rest mass. (It turns out that in relativity, a system's mass changes with its motion, so the invariant or rest mass is the portion of the mass of the system which does not depend on the system's motion.) One way of ensuring this energy-momentum relation is obtained is by substituting the Schrödinger equation as the evolution for the wave function with either the Dirac or the Klein-Gordon equation. Notice that the two equations are not equivalent: the former turns out to be suitable to describe particles with semi-integer spin (e.g., ½, fermions like electrons), the latter for integer spin (e.g., 1, bosons like photons). (Incidentally, as anticipated in Chapter 2, Schrödinger originally proposed the Klein-Gordon equation—before Klein and Gordon did—rather than the equation he is known for, as the evolution law for the wave function. He, however, dismissed it because it produced negative energy states.)

In addition to the Lorentz invariance of the laws, there are other crucial features of relativity that people would like to preserve also in quantum theory. These features, which arguably constitute the "spirit of relativity," are what people disagree about. As anticipated in Chapter 1, relativity was introduced by Einstein to reconcile electromagnetism—which predicted that light travels at constant velocity in all inertial frames of reference—with Newtonian mechanics, which is invariant under Galilean transformation and therefore predicts that velocity changes suitably when changing frames. Physicists assumed that light traveled through a medium called the luminiferous ether. However, the famous 1897 Michelson-Morley experiment failed to detect Earth's motion through the ether, suggesting that the ether might not exist. Einstein decided to eliminate the ether entirely and to rethink the concepts of space and time by proposing *two postulates: (1) the laws of physics are the same in all inertial frames and (2) the speed of light is constant in all inertial reference frames*. From these two, one can derive length contraction and time dilation: the clocks of someone moving relative to me tick more slowly than mine do; objects in motion are shorter in the direction of motion, from the point of view of a stationary observer. This theory, as you can see, is a *principle theory*; indeed, this is the context in which this distinction was proposed by Einstein. He was unsatisfied with this formulation and wanted to provide a constructive foundation to his own theory. A *historical constructive theory is the Lorentz-Poincaré ether theory*. This theory assumes that there exists a preferred frame of reference—the ether—through which light propagates. So, light's speed is truly constant only relative to the ether, and motion through the ether affects physical systems, thereby explaining length contraction and time dilation constructively. However, the role of spacetime in this theory is very different from the one in special relativity. *Arguably, a better constructive counterpart* of the original formulation of relativity is thinking of relativity as *a theory of the geometry of spacetime* based on Minkowski's interpretation. Instead of objects moving through space in time, objects exist in a four-dimensional spacetime structure where time and space are just different aspects of a unified whole. The spatiotemporal "length," called the interval and defined as $s^2 = c^2\Delta t^2 - \Delta r^2$, is invariant under Lorentz transformations, which are the transformations between inertial reference frames moving at constant velocities relative to each other. From this, it follows that relative motion affects dimensions and durations, leading to time dilation and length contraction, which therefore are not mechanical effects but consequences of the geometry of Minkowski spacetime. If relativity is a theory of spacetime, then it seems straightforward that for a theory to be relativistic, it has to be in *spacetime*, given that this is what relativity is about. If so, then high-dimensional approaches such as wave function realism are in a difficult position, given that they have no fundamental spacetime—even if one may argue that they have it nonfundamentally (see Chapter 3).

Otherwise—and this is likely what Einstein himself thought—*locality* is also crucial for relativity. As already discussed, the idea of locality is that physical influences propagate continuously through distances at finite velocity, and arguably, it has always been a *desideratum* of fundamental physical theories. Think, for instance, about the objections to Newton's theory of gravitation: Newton agreed it was a problem and replied that his theory was incomplete, and that a future better theory should make this action at a distance go away. In any case, the special theory of relativity poses an additional constraint on this idea—namely, that physical influences travel slower than the velocity of light, because in the theory *the velocity of light provides an upper limit for everything.* If so, influences cannot be instantaneous. Namely, there cannot be nonlocal interactions. The concept of fields, as an entity mediating the interaction by propagating through space at finite velocity, was introduced to get around the problem of instantaneous interaction. However, arguably, we know that quantum theories must be nonlocal, due to the violation of Bell's inequality. Does that mean that relativity is false? There is a sense in which this is the case: it is an empirical fact that locality must be abandoned.

Quantum correlations are *not information-bearing* because they cannot be used to transfer information faster than light; this is what is meant by saying they are nonsignaling. One might then wonder whether this is enough to reconcile quantum nonlocality with relativity. This was indeed Heisenberg's position (1949): quantum correlations are nonlocal in the sense that they travel faster than light, but since they do not transmit information, they do not conflict with relativity. However, someone who views relativity as a theory of spacetime—a constructive theory—will not be satisfied with this reply, as it effectively treats relativity as a mere principle theory.

Perhaps instead we should think of the spirit of relativity as being embodied in the spatiotemporal structure compatible with the theory. In this view, a theory that employs only *relativistic spatiotemporal structure* should count as genuinely relativistic. But what exactly is a relativistic spatiotemporal structure? In classical mechanics, we can talk about events happening in the same place, meaning they occur at zero spatial distance from one another. To give a silly example: if I'm sitting on a bench and you see, from another bench, a bird pooping on my head, then my location and the poop's location are at zero distance from each other, while yours is not. In other words, I am "equilocated" with the poop and you are not. Thus, equilocation is not absolute—it depends on one's point of view. Nonetheless, in classical mechanics, when we talk about events happening at the same time—meaning they occur at zero temporal distance—we mean this in absolute terms: my typing the letter "*t*" (event A) being simultaneous with Big Ben striking midnight (event B) is something everyone agrees on. This is because, in Galilean transformations, time is not mixed with position, unlike in Lorentz transformations. Because of this, in relativity there is no absolute fact about whether two events are simultaneous: *simultaneity is relative* to the observer's frame of reference. That means a *relativistic spacetime lacks a preferred frame*. In classical mechanics, one could "slice up" spacetime into objective simultaneity slices. In relativity, this is no longer warranted: there exists a foliation in which two events are simultaneous, but there is also another in which they are not and no one foliation is better than another. If so, one might say that a theory is relativistic if its formulation invokes no preferred frame or foliation of spacetime. Some have argued that at the microscopic level, pilot-wave theory is not Lorentz invariant, even though Lorentz invariance may emerge at the macroscopic level (Bohm and Hiley 1993, Holland 1993, Valentini 1997).

Alternatively, there have been *proposals for Lorentz-invariant versions of the pilot-wave theory*, using either the Dirac equation (Dürr et al. 1999) or the Klein-Gordon equation (Dürr et al. 1999, Nikolić 2005). However, all of these require a *preferred spatiotemporal foliation*—a specific slicing of spacetime. This is to be expected: due to nonlocality, the trajectory of any given particle is instantly influenced by the configuration of others. This necessitates an absolute temporal ordering—i.e., one must know where the other particles are *at the same time* as the particle of interest. That's what the preferred foliation provides. One can straightforwardly postulate such a preferred foliation (Berndl et al. 1996). Nonetheless, this strategy faces at least two concerns. The first is that such a foliation is *unobservable*: no one can empirically detect the preferred frame. The worry is that entities that are not, in principle, detectable are (arguably) irrelevant and should be eliminated. For instance, Galilean spacetime (which lacks absolute velocities, unlike Newtonian spacetime) arguably arises from the fact that absolute velocities are undetectable. However, in reply, one could argue that the case of the pilot-wave theory resembles the situation of absolute acceleration more than that of absolute velocity. Newton argued that absolute acceleration, though not directly detectable, is needed to explain observable effects (like the curved surface of water in a rotating bucket). Similarly, here, one might argue that we need a preferred foliation to account for relativistic effects. Its undetectability comes from our lack of access to particle configurations: if we could determine the exact trajectories of particles, we could also determine the preferred foliation. Thus, one could simply accept that there are aspects of physical reality that are inaccessible to us—especially when there's a principled explanation for why, as in this case. The second objection is that this addition seems *ad hoc*: it is introduced solely to "save the theory" rather than arising "naturally" from it. Some proposals have been put forward to respond to this challenge, the most promising of which suggests that the foliation can be suitably extracted from the wave function of the universe (Dürr et al. 2014). It is claimed that this makes the introduction of the foliation more natural, since all quantum theories already have such a foliation in virtue of having a wave function. However, one might reply that this doesn't necessarily entail a preferred frame: the fact that some mathematical object X (in this case, the foliation) can be extracted from the wave function does not imply that X has physical significance. Be that as it may, in fairness, the pilot-wave theorist can argue that the foliation does have clear physical significance: as mentioned, it is what is needed to account for nonlocal interactions. And, as noted, we can explain why it remains undetectable so there is nothing mysterious about it. For now, let's treat the preferred foliation as a feature of pilot-wave theory and explore the alternatives.

Beyond the Measurement Problem

It has been argued so far in this chapter that the pilot-wave theory is the obvious way of providing a constructive explanation which also preserves symmetries. The cost is, as just discussed, its nonlocality. So, the next question is: All other things being equal, can we get another viable quantum theory which is *more compatible with relativity than this?* The natural place to look is the other solutions to the measurement problem, after having checked whether they are compatible with a constructive type of explanation.

Spontaneous localization theories (GRW) and many-worlds (Everett) only contain the wave function, which, however, cannot be considered the fundamental ontology of matter in my approach, as it lives in a high-dimensional space. So, from my point of view, all these theories *are incomplete*, as they do not include anything specifying the fundamental nature of matter. The only option, then, is to specify some suitable fundamental ontology for them—one which is three-dimensional and foundational in the sense specified in the previous section. In other words, as usually presented, only the pilot-wave theory is satisfactory because it solves the measurement problem by completing standard quantum mechanics. In contrast, spontaneous localization theories avoid macroscopic superpositions by changing the evolution of the wave function, while many-worlds theories embrace macroscopic superpositions by inflating their metaphysics. Instead, within my framework, all quantum theories need completion, regardless of their solution to the measurement problem.

This means that, for me and for those who choose to endorse a view close to mine, the problem of making quantum theory compatible with realism is not the problem of eliminating macroscopic superpositions. In fact, two out of three solutions to the measurement problem (GRW and Everett) are unsatisfactory from my vantage point. Only completing the theory with some spatiotemporal ontology produces a satisfactory theory—otherwise, constructive explanation fails. So, it is hopefully now clear what *the problem of realism* is in this framework: *to make quantum theory compatible with realism, one needs to complete it with a foundational spatiotemporal ontology.* To put it another way, the problem to be solved in making quantum mechanics compatible with realism is not the measurement problem itself: some solutions to the measurement problem are not satisfactory because they are incompatible with constructive explanation. Having said that, the solution to the problem of realism is also straightforward: *drop the wave function* as an object representing matter and instead have something in spacetime representing material systems. If we do not have such a three-dimensional ontology, then we run into undesired (and unnecessary) difficulties, such as radical ontologies and convoluted explanations. The reason is the one we have just seen: only a three-dimensional, effectively microscopic, foundational ontology is compatible with a symmetry-preserving constructive picture. Consequently, a theory which solves the measurement problem by denying one of the three assumptions discussed in Chapter 2—the wave function being the complete description, the wave function evolving according to the Schrödinger equation, and the existence of macroscopic superpositions—is not necessarily going to be satisfactory. Rather, it is satisfactory only if it denies that the wave function represents the fundamental material ontology, and if it postulates the existence of some suitable three-dimensional entity evolving in time (or a four-dimensional spatiotemporal entity) to describe physical objects.

Which Ontologies?

As seen in the previous section, among the theories we have discussed so far, the only one that provides a constructive explanation is the pilot-wave theory. In fact, it can be naturally understood as a theory with a particle fundamental ontology, guided by a Schrödinger-evolving wave function. So, what about the other theories? It is possible, in principle, to *construct theories in the spirit of GRW and Everett that would fit the framework* discussed in this chapter by adding to each of them a suitable spatiotemporal dimensional ontology. But which ontologies would be suitable?

Before exploring this question, one may wonder *why there is even a choice* to be made in the quantum case, given that in classical mechanics this question never arose. The reason is that in classical mechanics, it was clear from the very beginning what the metaphysical hypothesis about the nature of matter was, while this is not the case in quantum theory. Newton started with the hypothesis that the world is made of particles and continued accordingly. Presumably, the law of evolution in classical mechanics was selected as the simplest among the (infinite) empirically adequate alternatives. Arguably, no one else considered any other alternative ontology because that would have been pointless: they already had a very successful theory, so why speculate? Also, waves as fundamental entities were ruled out by the lack of interference of matter. This did not happen in quantum theory, and its history also contributed to making it difficult to think about realism to begin with, let alone focus on a single ontology (see Chapter 7). As a result, different realist proposals emerged, each with their own difficulties, leading to the current pool of possible quantum ontologies.

We have seen how, in the case of a particle ontology, there is not much of a discontinuity with the classical way of thinking. A particle picture is one in which the world is discrete in space. Alternatively, one could have a wholly *continuous picture* at the fundamental level, in which matter is represented by a matter or charge density field (in spacetime). In this picture, one must ensure that the matter density combines in such a way as to give the impression of having particles at some nonfundamental level. That is, long-lasting and nonspreading *wave packets* need to form and persist, and arguably, decoherence—namely, the interaction with the environment—ensures that. Otherwise, one could have a fully discrete picture of *isolated spatiotemporal events*, or "flashes." They are isolated in the sense that they are not connected with one another by trajectories. However, each flash can be seen as a particle appearing and disappearing at one time and then reappearing at a different time. A succession of flashes would be the natural extension of a particle trajectory in the case of discrete time, even though a flash ontology is compatible with a continuous view of time. In this picture, the world is discrete in space in the sense that matter is not continuous, as our macroscopic perceptions suggest, but is made up of point-like particles. Additionally, the world is also discrete in time: these particles "flash" in and out of existence; they only live for an instant and reappear at a different instant. At a sufficiently large temporal scale, we can think of these flashes as connected by continuous trajectories, just as at a sufficiently larger scale, we think that a suitable group of particles forms a continuous body. So, even though the flashes might provide an unfamiliar ontology, dynamical and compositional analysis is still possible.

Various considerations lead to different proposals: simplicity leads to particles, similarity with electromagnetism leads to fields, and considerations about relativity lead to flashes, as we will discuss later in this chapter. Nonetheless, historically, people focused on the wave function (see Chapter 7). In fact, the various solutions to the measurement problem differ in the way the wave function evolves: in many-worlds theories, it evolves deterministically according to the Schrödinger equation, while in spontaneous collapse theories, it evolves according to an indeterministic, nonlinear modification of the Schrödinger dynamics. In the next section, we will see how to think about these theories from the perspective of having a spatiotemporal foundational ontology.

Many-Worlds Theories?

We can complete the other solutions of the measurement problem with a spatiotemporal ontology. But what would these theories look like? Let's begin with many-worlds theories. By definition, they describe reality as constituted by an infinite multiplicity of effectively noninteracting and, therefore, inaccessible "worlds," each depicting another way the actual world could have been. As I previously argued, the historical first example of a many-worlds-type theory was provided by Schrödinger himself, who briefly considered treating the square of the wave function as a *charge density* (1926). This is an example of a field ontology, in which matter is understood as fundamentally continuous. Schrödinger considered identifying this field with charge because he thought he could directly connect quantum theory with electrodynamics. However, Schrödinger soon abandoned the field interpretation for two reasons. As mentioned earlier, the first is that the wave function is defined in configuration space. This can be avoided by "integrating out" all the degrees of freedom of the wave function except three, thus resulting in a three-dimensional charge density. Alternatively, one could identify such a quantity with mass, rather than charge. In any case, in this interpretation, what we commonly call particles are nonfundamental, as they are identified as wave packets. However, aside from the fact that the wave packet quickly spreads, this theory still faces the measurement problem: macroscopic superpositions of matter density fields inherit the linear evolution of the wave function. As a result, the theory is empirically inadequate, and Schrödinger dismissed it. In fact, he even wrote his famous cat paper to demonstrate how absurd it was (1935a). Nonetheless, contrary to Schrödinger's dismissal, *one can explore the consequence of thinking of this theory as really depicting a multiplicity of noninteracting, thus inaccessible, "worlds."* A theory such as this, in which the fundamental three-dimensional ontology is a matter field, has been labeled *Sm*: *m* denotes the material ontology, and *S* indicates that the evolution of the matter density is governed by a Schrödinger-evolving wave function (Allori et al. 2008). Everything we've said about Everett mechanics also applies here, with the addition that matter exists in spacetime. In fact, the various terms of the superposition in the matter density field can be thought of as effectively independent: due to their interaction with the environment (decoherence), they do not interact with one another. Since the alive cat and the dead cat do not interact with each other, they can be seen as corresponding to alternative states of the cat. Thus, they describe alternative worlds superimposed on a single spacetime. Since the different worlds do not interact among themselves, they are, so to speak, reciprocally transparent: they do not "see" one another. Other theories with a many-worlds character have been proposed. A *particle many-worlds* theory was originally discussed by Bell (1987) and is sometimes dubbed BMW ("Bell Many Worlds," Allori et al. 2008). (This has also been called *Sip* in Allori et al. 2011. In this notation, *Sp* would refer to the pilot-wave theory: *p* for particle ontology and *S* for Schrödinger wavefunction evolution. However, here there is an "i" for "independent" to indicate that BMI has an ontology of instantaneous, randomly distributed particles with no temporal correlation among them, whose probability distribution is governed by a Schrödinger-evolving wave function.) Another theory, dubbed *Sf* (Allori et al. 2008), has an ontology of *flashes*, whose distribution is determined by a wave function evolving according to Schrödinger equation. These theories have a many-worlds character because, since there is no connection between different configurations at different times, the configuration will likely visit distant regions at subsequent instances. That means, for instance, that if there are dinosaurs at time t, they will have disappeared by time $t + dt$. Therefore, many worlds exist, not at the same time, but one after another (for more on these theories, see Allori et al. 2008, 2011).

If one were looking to construct theories that depict many worlds, there are presumably other ways to do it. However, there seems to be *little gain in even considering them*, at least from my constructive perspective. Indeed, one should wonder why, in the quantum domain, many-worlds theories are even considered (see Chapter 7). After all, in the pre-quantum era, no one ever thought about them as a viable possibility, and for good reasons: *if a theory predicts macroscopic superpositions that are not observed, the most natural thing to say is that the theory has been falsified*—meaning the proposal is not empirically viable. Arguably, in the past, theories that required accepting the existence of many worlds to account for the prediction of unobserved macroscopic superpositions would have been immediately discarded. This is precisely what Schrödinger did: he abandoned such a theory. Nonetheless, this attitude had nothing to do with his being a realist. Indeed, good empiricists, who value what they can observe, should never have accepted a theory that proposes an assumption just to avoid falsification. Arguably, this is exactly why Schrödinger discussed the problem of observable macroscopic superpositions: if empiricists can ignore unobservable superpositions because they cannot directly access them, they still have to care about the observable ones. Moreover, the very hypothesis that these theories propose—namely, accepting the existence of noninteracting, and therefore unobservable, worlds—should be considered *preposterous* from the empiricist's point of view. The fact that physicists, who are usually more empiricist than the typical metaphysician, accepted the many-worlds hypothesis, as I will underline in Chapter 7, has to do with a range of factors, none of which seem to have much physical merit. If that is the case, many-worlds theories seem to be *the last resort for those who do not want to admit that their theory is empirically inadequate*. Therefore, it is not a problem about an inflated metaphysics, and it does not concern super-empirical virtues, about which people can arguably disagree. Rather, accepting many-worlds theories is a matter of empirical adequacy—something on which everyone agrees, from the hard-core realist to the empiricist. Everyone agrees that a viable theory needs to at least reproduce observations, which many-worlds theories do only by accepting the existence of an infinity of in-practice undetectable worlds. There are infinite ways of modifying a theory to make it cope with observation, one of which is to postulate that we do not observe the other superpositions because they belong to another "world" that is inaccessible to us. In other words, adopting a many-worlds metaphysics just looks like *an instance of the Quine-Duhem thesis*: if you have enough imagination, you can always save what you want from falsification, but that does not mean it is the reasonable thing to do. As I argue in Chapter 7, had history taken another course, no one would have taken this tactic seriously. But things took a different turn, and theories that escape falsification by postulating unobservable worlds are currently considered serious contenders.

On another note, Let me conclude by observing that some many-worlds supporters will argue that many-worlds theory is worth taking seriously because, even if it is not separable, it is the only *local theory*, and we should value locality for the reasons we have seen (Wallace 2012). Nonetheless, as already discussed in Chapter 2, this is *controversial*: didn't Bell's theorem show that quantum mechanics is nonlocal, regardless of the details? The fact that many-worlds theories violate Bell's assumption that there are unique experimental outcomes may be the crucial observation. Nonetheless, it has been argued that, even under these constraints, many-worlds theories are nonlocal (Allori et al. 2011). In short, one can show that even if there is no actual interaction between the physical objects, there is nonlocality within the same "world." In fact, consider *Sm* and an EPR-type situation, in which, say, a *z*-spin measurement is performed. Suppressing the two components, dropping the numerical weights, and denoting $\sigma^{+}\sigma^{-}=\sigma^{+,-}$, after the measurement, the wave function will be in a superposition of the form $\sigma^{+,-} + \sigma^{-,+}$, since there is no collapse, and so it will correspond to the matter density function $m_{+,-} + m_{-,+}$, where each term corresponds to a possible world. This means that the world in which the result for particle 1 is "up" (+) is the same world in which the result for particle 2 is "down" (−). In any case, even if it turns out that the world is, in this sense, local within a many-worlds perspective, would saving the intuitive notion of locality be worth the extravagant, empirically ungrounded metaphysics, even considering the mathematical simplicity of the theory?

Spontaneous Collapse Theories?

Now, let's turn to the last solution to the measurement problem, namely the GRW-type theories, characterized by their nonlinear stochastic wave function evolution. The first historical proposal of a spatiotemporal fundamental ontology is GRWm, where "*m*" stands for a matter density field, and it is guided by a GRW-evolving wave function. In this theory, matter is conceptualized as a three-dimensional field representing a *mass density field*, defined as the wave function density integrated over all but three dimensions, which stochastically and nonlinearly evolve as prescribed by the GRW dynamics (Benatti et al. 1995). (Presumably, the proponents of this ontology chose mass over charge, as Schrödinger suggested, because they considered mass to be a more fundamental property of matter than charge. This represents Schrödinger's perspective in a theory in which there are no many worlds, as macroscopic objects quickly collapse.) What we call "particles," understood here as nonfundamental matter-density packets, are suitably localized. Effectively, for macroscopic objects, all the "worlds" are collapsed by the GRW evolution. For a fraction of a second, macroscopic objects can exist in superpositions, but then all but one of the superposition terms disappear due to collapse.

Another proposal for a spatiotemporal fundamental ontology for a GRW-evolving wave function is one based on *flashes*, defined as the spatiotemporal points where the wave function localizes. In this theory, the collapses of the wave function identify where the flashes are located in spacetime. This theory, GRWf, was proposed by Bell (1987).

Alternatively, one could insist on maintaining the simplest type of ontology, namely *particles*, even for stochastic localization theories. Such a theory has been proposed under the obvious name of GRWp (Allori et al. 2011, Allori 2020b). It turns out that for GRWp to be empirically adequate, one would need a stochastic evolution for the particles as well: the particles follow a guidance equation, as in the pilot-wave theory, but when the wave function jumps, they are displaced at random.

All these theories are empirically viable (they have no unobserved macroscopic superpositions), and they may all provide a constructive explanation of phenomena because they have a spatiotemporal ontology. Moreover, they respect symmetries, as in all these theories, the ontology is foundational, and the wave function is nomological. Thus, the wave function can transform as needed to ensure that the law for the evolution of matter also transforms to maintain symmetries. This means they all seem to be equally valid alternatives to the pilot-wave theory. Nonetheless, it seems reasonable to choose the *simplest ontology* (particles) and the *simplest type of law* (linear and deterministic), namely the pilot-wave theory, especially within my framework, in which all quantum theories need to be completed with a spatiotemporal ontology (Allori 2019b). So *why look for more*? Indeed, particles could be considered the simplest ontology: they require just one parameter to be specified—their location in space. Additionally, particles are more familiar, as previous, well-developed, and well-known theories like classical mechanics have had a particle ontology. Again, simplicity and familiarity are not necessarily truth-conducive, but all things being equal, there seems to be no reason to endorse theories with unnecessarily complex or unfamiliar ontologies. Moreover, as discussed earlier, *closeness to the manifest image should be preferred*, especially considering that a particle ontology would defeat the pessimistic meta-induction argument. If that is the case, and we should privilege a particle ontology, then the pilot-wave theory stands out. It is not the only theory with a particle ontology, but there does not seem to be any point in considering theories like GRWp or BMW: the former involves a stochastic wave function and particle evolution, which is unnecessarily more complex than the deterministic one in the pilot-wave theory, while in the latter, particle evolution is uncorrelated, which implies a many-worlds character.

Arguably, it is difficult to see any advantage in opting for a matter density ontology over a particle ontology. In fact, Sm is arguably better than GRWm, as the equation for the wave function is simpler in the former than in the latter. However, as we have seen, Sm has a many-worlds character, and choosing such an extravagant metaphysics—which is neither natural nor mandatory—is difficult to support in the absence of independent justification.

Flashes are undoubtedly a more radical choice, as they prescribe that matter is very different from our everyday perception of it: discrete both in space and in time. Thus far, there seems to be no compelling reason to consider them. If we set this concern aside for the moment, among the theories with flashes, GRWf appears to offer the best balance of mathematical simplicity and metaphysical sensibleness. This is because it does not possess a many-worlds character like Sf, even though, to avoid that, we must have a stochastic nonlinear evolution for the wave function.

As previously mentioned, in my framework, the only valid reason to consider theories other than the pilot-wave theory relates to other criteria of theory choice, namely *symmetries and fruitfulness*. In particular, it has been noted that the flash ontology of GRW is closely connected to *relativistic invariance* (Bell 1987). GRWf is "as relativistic invariant as it could be," displaying relative time-translation invariance, which is considered the nonrelativistic "shadow" of Lorentz invariance. A theory is time-translation invariant if two distant systems can be shifted in time by different amounts. This assumes that absolute simultaneity plays no role in the theory: if it did, the times in the two distant systems could not be shifted independently. GRWf is relative time-translation invariant, while GRWm is not. In fact, according to GRWf, in an EPR-type experiment, each flash is a spacetime point that correlates with nothing on the other side. There is no time ordering between the events in the two wings; instead, there is only a probability distribution of flashes, which does not depend on whether the events were simultaneous. In contrast, in GRWm, absolute simultaneity can be inferred from the matter density: the moment when the matter density disappears in one location is the same moment at which the detection occurs on the other side. Inspired by this observation, the first successful attempt at a relativistic GRW-type theory, rGRWf (where the "r" stands for "relativistic" and "f" for "flashes"), was proposed by Tumulka (2006a, 2021). This theory is *Lorentz invariant and has no preferred foliation*. It specifies the joint probability distribution of all future flashes, where the initial conditions include a wave function defined on an arbitrary spacetime slicing and a "seed flash" for each "particle." Instead of universal time, the theory uses hyperboloids foliating the interior of the light-cone of a seed flash, which assigns a proper time to each event. (The light cone is the path that light emanating from a point would take through spacetime; the past light cone of an event consists of all the events that could potentially have influenced the given event.) The hyperboloid structure also defines the collapse mechanics, as one specifies a Gaussian function relative to it. This theory is nonlocal: for two "particles" in an entangled state, the collapse associated with a flash on one particle can change the probability distribution for the flashes of the other particle, regardless of their space-like separation.

For a while, people thought that flashes had something special that might make a flash theory more compatible with relativity. However, more recently, a *Lorentz invariant collapse theory with a matter density ontology*, rGRWm, has been proposed. In this theory, the matter density is defined at each point with respect to the past light-cone at that point (Bedingham et al. 2014). Therefore, rGRWm is relative time-translation invariant, Lorentz invariant, and foliation-free. This theory is also nonlocal for similar reasons as rGRWf. Given that a matter density ontology is less radical than the flash ontology, one should prefer rGRWm to rGRWf. Additionally, there seems to be no in-principle obstacle to having an rGRWp. If one could find such a theory, *rGRWp should be preferred*, all other things being equal, due to its particle ontology. I do believe it is the best option, even though some argue that a particle ontology is in tension with relativity (Myrvold 2021). This issue requires further exploration, but the main concern is that relativistic quantum theories should be theories of fields rather than particles. However, it is not obvious that quantum field theories are actually theories of fields, as some quantum fields theories with a particle ontology have been proposed. For this reason, this concern seems unfounded (Dürr et al. 2004, 2005b).

I have recently argued that the *key factor* allowing GRW theories to be Lorentz invariant and foliation-free is their *stochasticity* (Allori 2022d). In fact, it is not the nature of the fundamental ontology of matter that is relevant. Lorentz invariant GRW theories can be provided for multiple ontologies. In fact, all proposed Lorentz invariant GRW-type theories are "doubly" stochastic: both in the evolution of the wave function and in the evolution of the fundamental ontology of matter. One cannot dispense with this double stochasticity because a wave function evolving stochastically must combine with a fundamental ontology that evolves stochastically. Flashes are intrinsically stochastic; the matter density in GRWm inherits the stochasticity of its evolution from the stochasticity of the wave function's evolution, while the particles in GRWp must evolve stochastically for the theory to be empirically adequate. One could attempt to use a deterministic wave function evolution and a stochastic evolution for matter, but it turns out that such a theory would still require a foliation-like structure. In fact, it can be proven that any deterministic theory must have a preferred foliation (Ghirardi and Grassi 1996). If this is the case, then only stochastic theories can be *Lorentz invariant and foliation-free.*

Assessment

Given their exclusive reliance on relativistic spatiotemporal structure, Lorentz invariant GRW-type theories initially seemed to align better with relativity and were thus worth investigating. This was, at least, my initial view. However, I have since come to believe that this advantage is not worth the costs when compared with relativistic pilot-wave theories.

First, their *nonlinearity and stochasticity are not ideal.* These theories are nonlinear and stochastic because theorems have shown that linear stochastic theories allow superluminal signaling and, therefore, must be discarded (Gisin 1984). Whether this conclusion is correct or not, their stochastic nature already makes them more complex than their pilot-wave counterpart. Apart from the nonlinearity, deterministic laws seem better in any case, as they straightforwardly allow for constructive explanations. The situation with indeterministic laws is less clear: it's not obvious that stochastic laws allow for the same type of explanation. For instance, due to their stochasticity, the notion of cause and effect—and, consequently, the notion of causal explanation—becomes problematic (Allori 2022d). Consider, for example, an EPR-type experiment. In GRW-type theories, *cause and effect are symmetrical.* When there are no continuous trajectories (as with a flash ontology, which is discrete in both time and space, or with GRWm and GRWp, where matter jumps), it's difficult to say in what sense one event causes another. This directly violates the principle of uniformity of causation, which states that given a cause, a given effect will certainly follow. This creates a tension between these theories and the possibility of a causal explanation. One might still argue that causation is not necessary for a constructive explanation if causation is not considered fundamental. All that is needed is the possibility of thinking about macroscopic objects being suitably composed, with their properties dynamically reduced. Indeed, it is plausible that these theories, whether stochastic or not, allow for constructive explanation: on a suitable scale, their ontology, provided it is spatiotemporal, will behave as if it were localized, and the randomness of their trajectories will smooth out. This is a requirement if these theories are to reproduce, in some limit, classical mechanics, which involves particles moving along continuous trajectories. That said, more importantly, the fact that in GRW-type theories the notions of cause and effect become symmetric *threatens the very idea of influence*, which presupposes these notions. One might argue that there is still room for the notion of cause and effect, at least in a frame-relative sense. For example, in an EPR experiment where Alice measures her particle first, her action could be seen as the cause of the result on Bob's side. However, there are frames where one event is the cause, and frames where the same event is the effect. Since there is no privileged frame, these frames are equally legitimate, and thus cause and effect lose any absolute meaning. Moreover, it seems that even if we could save the notion of causation or influence, we would still encounter superluminal influence or "nonlocal probabilistic correlation" (Myrvold 2016, 2021). Arguably, in GRW-type theories, the interaction notion of locality—the idea that influence travels at a finite velocity—seems to have evaporated.

Another related problem is that these theories exhibit nonlocal interaction within a single system, a feature I dubbed "*supernonlocality*," which renders the notion of influence meaningless (Bricmont 2016, Allori 2022d). Consider an experiment with a single "particle" confined in a box, viewed from the perspective of rGRWm. Suppose someone inserts a barrier in the box, splitting the wave function of the "particle" into two. The two half-boxes are then separated, at which point one of them is opened. In rGRWm, since there is only one "particle," no wave function collapse has occurred. The matter density is distributed equally between the two half-boxes. Now, suppose someone opens, say, the half-box on the right. The "particle" couples with the measuring device, and the wave function collapses. That is, the matter density jumps into one of the half-boxes. Since the process is random, the matter in the right half-box might jump to the left (and nothing is found when opening the right half-box), or it might all jump to the right (and the "particle" is found). In either case, there is nonlocal matter transfer even with just one "particle." Nothing like this happens in the pilot-wave theory, hence the term supernonlocality. To observe nonlocal effects, one usually needs at least two distinct systems, but here, only one is necessary. The case for the other ontologies is similar, as the stochasticity in their evolution is what underlies supernonlocality. One might respond that these examples do not demonstrate true nonlocal matter transfer. One could liken them to shining a laser pointer across the face of the Moon: the spot can move faster than light, but the "spot" is not a tangible entity moving (Norsen p.c.). However, this analogy is not applicable to these theories, as it would imply that the matter density, the flashes, and the particles are not really things that move, which contradicts the approach I advocate. Another possible reply is that supernonlocality does not worsen the compatibility of these theories with relativity. In fact, one could argue that either a theory is compatible with relativity or it is not, and the number of times a theory exhibits incompatibility does not matter. Both theories are nonlocal, and the fact that a supernonlocal theory shows nonlocality in more situations does not make it worse. However, I would counter that supernonlocality is unnecessary (as shown in the case of pilot-wave theories), and it does not seem to be good scientific practice to favor theories that are more radical than necessary. Additionally, to reiterate, nonlocality pertains to composed systems: when you have two or more systems, they influence one another nonlocally. However, *since supernonlocality occurs in a single system, it is unclear what the concept of locality even means in these theories.*

Another thing to notice is that GRW-type theories are *not empirically equivalent* to quantum mechanics because the Schrödinger evolution no longer holds. For instance, they predict that energy is not conserved. Even if this can be theoretically explained, it remains difficult for many to accept. One might think that, unlike the pilot-wave theory, at least these theories are testable (in the simple sense that it makes different predictions from quantum theory). However, they contain parameters (such as the localization accuracy and the collapse frequency) that can be adjusted to avoid falsification: whenever an experiment confirms the quantum predictions, GRW-type theories can also be confirmed by tweaking these parameters. This makes the theory highly *ad hoc* (Bricmont 2016).

The lesson I draw from all this is that these theories show the consequences of insisting that relativity is a complete theory of spatiotemporal structure. In other words, quantum theory is a theory of nonlocal interaction, while relativity theory describes Lorentz-invariant spatiotemporal structure. This structure contrasts with nonlocality, which seems to require a preferred foliation. GRW-type theories heavily modify quantum theory with a nonlinear stochastic evolution that generates new, unwelcome predictions. These theories also contain parameters that can be freely adjusted to avoid falsification, and they cause the notions of cause and effect, as well as influence, to disappear. One must accept all of this in order to maintain relativity as it is. In contrast, *the pilot-wave theory does the obvious thing: it minimally modifies standard quantum mechanics by adding a spatiotemporal ontology (particles) and minimally modifies relativity by specifying a preferred foliation on which nonlocality acts.* This nonlocality is in tension with relativity, which is fundamentally a local theory (Maudlin 1994). However, this tension reflects an experimental fact that cannot be dismissed. Thus, it seems more of a merit than a problem that in the pilot-wave theory, *nonlocality* is as explicit as possible, and we should interpret it as evidence that relativity is incomplete. That is, it *provides evidence for the existence of a fundamental preferred frame*. Keep in mind that this theory does not break Lorentz invariance: laws still appear the same under all inertial frames, provided we also assume the constancy of the velocity of light. What no longer holds is the relativity of simultaneity.

Given these considerations, I am now convinced that the pilot-wave theory is not only the rational completion of quantum mechanics (as Jean Bricmont often argues), but it also represents the *rational completion of relativity* (see Allori 2024b for more): *quantum mechanics is incomplete because it does not have a spatiotemporal otology, and relativity is incomplete because it does not accommodate quantum nonlocality.* The pilot-wave theory minimally modifies both to make them compatible: it adds particles to quantum theory and the preferred foliation to relativity. As Detlef Durr used to say (as a general claim in favor of the pilot-wave theory, p.c.), "this is just what the doctor ordered."

Be that as it may, if we wish to endorse the pilot-wave theory, we must better explain what the wave function is within this framework, and that will be the focus of the next chapter.

Further Readings and Discussion

The core idea behind the primitive ontology approach, which underpins my account, is succinctly articulated in Dürr et al. (1992) and later expanded upon in Allori et al. (2008). I have explored and defended this perspective in numerous articles, both individually (Allori 2013a, b, 2015a, b, 2018a, b, 2019b, 2020b, 2021a, b, 2023a, b) and in collaboration with others (Allori et al. 2008, 2011, 2014).

However, I had not delved into the details of how classicality emerges in the pilot-wave theory or the role of decoherence in this process. Articles discussing the classical limit include Bohm (1952), Bohm and Hiley (1993), Holland (1996), Allori et al. (2002), Bowman (2005), Allori and Zanghì (2009), Rosaler (2015), and Allori (2016). For more on decoherence in quantum mechanics, see Bacciagaluppi (2020), Crull (2013), Romano and Vassallo (2021), and the references therein.

Further discussion on no-go theorems in relation to the pilot-wave theory and the EPR experiment can be found in Hemmick (1996), Hemmick and Shakur (2012), and more recently in Bricmont (2016), Bricmont et al. (2020a, b, 2022).

On the topic of nonlocality, in addition to the references provided in Chapter 1, I recommend the works of Norsen (2006a, b, 2009, 2011, 2015, 2016). For the distinction between parameter and outcome independence, which I have not mentioned, see the additional papers mentioned in the text, as well as Redhead (1987), Cushing and McMullin (1989), Butterfield (1992), Ghirardi (2012), and Myrvold (2016).

Regarding relativity, relativistic invariance, and space-time structure, useful books include Anderson (1967) and Friedman (1983), as well as DiSalle (2020). For more general concerns about the intersection of relativity and quantum theory, see Myrvold (2021). In particular, for the relationship between relativity and the pilot-wave theory, refer to Dürr et al. (2014) and the references therein. For relativistic spontaneous collapse theories, see Tumulka (2006a), Myrvold (2019, 2021), and Bassi and Ghirardi (2020), in addition to the references in the main text.

5

Demystifying the Wave Function

In the previous chapter, I presented my preferred approach, particularly discussing how one should proceed in choosing a theory in cases of underdetermination: by selecting *the simplest theory that allows for a constructive explanation while preserving symmetries*. These considerations naturally lead to the pilot-wave theory, understood as a theory of particles guided by a wave function. However, it remains unclear how to understand the wave function within this theory (and more generally, within any quantum theory that is compatible with this approach).

While the last chapter focused on the fundamental ontology of matter, this one is devoted to the wave function. Here, I present a novel approach to understanding the nature of the wave function in theories with a spatiotemporal material ontology. The basic idea is that the wave function (is part of what) describes the interaction between the fundamental entities, rather than the entities themselves. Nonetheless, strictly speaking, it does not represent the interaction. That is, it should not be understood representationally, but functionally, in terms of the empirical, explanatory, and physical roles it plays within the theory. Accordingly, I have adopted the term *wave-functionalism*, a label proposed to me by Mario Hubert, to describe my view. I conclude this chapter by noting how a suitable extension of *explanationism*, as presented in Chapter 1, pairs naturally with this view. According to this type of realism, one is justified in being realist about the entities responsible for the theory's success, which I refer to as the working posits. Since, in the case of quantum theory, the wave function is not a working posit, one can remain "metaphysically neutral" about its status: it is not necessary to assume the wave function exists in order to make sense of quantum theory from a realist perspective.

Chapter Schema Details

In the first section, I provide an *overview of the motivations and the problems* associated with various views on the nature of the wave function when thinking of it as nonmaterial, like in my approach and others. The fundamental question is: *If the wave function is not a material field, what is it?* Nomological approaches suggest identifying the wave function with laws, properties, or, more generally, structure. Pragmatist views regard the wave function as prescriptive, guiding our beliefs about experimental outcomes. Epistemic approaches treat the wave function as expressing our ignorance about the state of the system.

In the second section of this chapter, I present my account of the wave function—wave-functionalism—which holds that *the wave function is defined by the roles it plays within the theory*. That is, it should be understood in terms of its functions: enabling empirical predictions, explaining observed outcomes, and preserving the theory's symmetry properties. Because it is functionally defined, the roles the wave function performs can be multiply realized. To motivate my view, I argue that it combines the strengths of the alternative approaches while avoiding their shortcomings. In contrast to nomological approaches, the wave function can be time-dependent and nonunique. In contrast to epistemic approaches, it can account for interference phenomena. And in contrast to pragmatist approaches, it is compatible with a compositional and dynamical understanding of explanation.

This leads us to the third section, where I discuss *the type of realism that naturally fits* with this approach. This realism is a suitable extension of what I referred to as *explanationism* in Chapter 1. The idea is relatively straightforward: we are justified in believing in the existence of what we have identified as the fundamental ontology of matter, which serves as the working posit of the theory, while *the wave function is merely a presuppositional posit*. This means that we do not need to be realist about the wave function; it is just an idle part of the theory. What is preserved in the transition from classical to quantum theory is the fundamental ontology of matter, not the wave function. Therefore, we can remain metaphysically neutral about the wave function without considering it epistemic: the wave function is objective (i.e., mind-independent) but does not represent any objective fact. This perspective avoids the shortcomings of the original version of explanationism, and offers a robust response to the pessimistic meta-induction argument against scientific realism, discussed in Chapter 1: one should focus only on the fundamental material ontology, not the wave function, to determine whether it carries over during theory change. The particle ontology of the pilot-wave theory is ideal in this respect, as it is identical to the one of classical mechanics.

The Mysterious Wave Function

As discussed, from my perspective, the wave function does not represent some new and peculiar material field that needs to be added to "the furniture of the world," as its high-dimensional nature prevents a constructive explanation. It should also not be considered as representing a three-dimensional multi-field or anything similar to what other approaches propose (e.g., material objects, exclusively local beables, primitivism, or multi-field theories), as this would cause the theory to lose its symmetry properties. So, if the wave function does not represent matter, what is it? Clearly, the wave function plays several central roles in the theory: there is an equation of motion for it, and the square of the wave function provides the probabilities of experimental results. But what do these facts tell us about the nature of the wave function? Several proposals have been put forward to explain what the wave function is, given that it does not represent a material quantum state. Therefore, the question for this chapter is as follows: *What is the best way to understand the wave function, if it does not represent a material field?* There are three major types of alternatives: nomological accounts, epistemic accounts, and pragmatist accounts. Within these classifications, there are several different views, so let's explore them in more detail.

The Wave Function as a Law

The idea behind nomological approaches is that the wave function represents a nomological fact, a law of nature. This is the traditional way of understanding the wave function within the original primitive ontology framework: the primitive ontology specifies what matter is, and the wave function tells matter how to move (Allori et al. 2008). After all, the wave function is not at the foundation of the theory's architecture, and as we've seen, these entities are better understood as nomological. This general idea can be articulated in different ways, depending on what is meant by the term "nomological." One possibility is to take the suggestion literally: *the wave function is a law of nature*. However, this can't be entirely correct, as the wave function does not resemble a law. First, intuitively, laws are universal generalizations. That means they are immutable and constant, time-independent. By contrast, the wave function *changes over time*: it evolves according to the Schrödinger equation. What kind of law changes over time? Evolution equations are meant for objects, not for other things—especially not for laws. The very idea of a law implies that it is timelessly valid.

The usual response here is defensive: there are promising indications that, in a future theory of gravity, the wave function will be static and thus more law-like (Goldstein and Teufel 2001, Goldstein and Zanghì 2013). However, we would like to have a good account of the theory we currently possess, rather than wait for a future theory. Otherwise, one risks falling into anti-realism. A more plausible reply might be to resist the assumption that laws must be constant and immutable. True, this is our intuitive understanding, and it's how things have been thus far, but science is telling us that some laws change, and we should update our conception of what laws are (Callender 2015). This reasoning seems natural from a Humean perspective on laws, where laws are simply generalizations that supervene on the Humean mosaic. However, accommodating this idea is much more challenging from an anti-Humean perspective.

Another related problem is that our intuitive notion of laws suggests that there is only one law. However, *the wave function is not unique*: as we have seen, two wave functions differing only by a phase produce the same empirical outcomes. If the wave function is a law and it is not unique, then the law itself would not be unique, which seems to contradict our intuitive understanding of what laws are supposed to be.

One might respond to this challenge along the lines discussed earlier. Alternatively, one could argue that there could be different formulations of the same law, expressed in terms of different mathematical objects. For example, in classical mechanics, the same law can be written in terms of a potential V or $V+c$: similarly, one could write the same law in terms of the wave function ψ or $\psi e^{i\theta}$. However, some may question this analogy, pointing out disanalogies between the wave function and the potential function (Brown and Wallace 2005). If these criticisms hold, one would not be able to claim that two wave functions differing by a phase would represent the same law.

Regardless, the key point of these objections is that thinking of the wave function as "the law" seems to be a *category mistake*: laws govern (or describe) the behavior of matter, while the wave function itself is governed by a law.

The Wave Function as a Property

Alternative approaches to understanding the wave function suggest that it is related to the way matter behaves, rather than to matter itself, and that it should be considered in terms of properties rather than laws. In this view, *the wave function represents a property of matter*, which explains (or accounts for, or describes) how matter behaves. In classical mechanics, properties like mass or charge are typically considered intrinsic to matter, and these properties account for why matter moves in certain ways. For instance, an electron moving through a magnetic field will be deflected "up," while a proton will be deflected "down," because they have opposite charges. Similarly, one might argue that the wave function plays a comparable role: it is because an electron has an anti-symmetric wave function, rather than a symmetric one, that it follows the Pauli exclusion principle in atomic orbitals. In quantum mechanics, however, the situation is complicated by phenomena like entanglement. As discussed in Chapter 1, some property-oriented approaches to scientific realism aim to generalize this strategy. For example, consider a pair of entangled particles in a singlet state. According to these approaches, the entanglement between the particles is a relation that is not explained by any individual property of the particles. There is nothing intrinsic to a single particle that determines their mutual relation, which is described by the wave function. The entangled wave function thus expresses a brute, irreducible fact about the particles' relation. "Having opposite spins in that direction" is a property, but it is not a property of the individual particles; rather, it is a property of the entangled pair. This property is irreducible because it is not determined by more fundamental properties of the individual particles (Monton 2002, Suárez 2015).

That said, this approach still faces the problems outlined earlier. Fundamental properties, like mass, are expected to be *unique and time-independent*. For example, if mass is considered a defining property of a particle (e.g., a particle is an electron because it has a mass of $= 9.1093837 \times 10^{-31} kg$), it should be unique and constant over time. Therefore, to move forward with this approach, one would need to treat *the wave function as a dispositional property*. As discussed more thoroughly in Chapter 6, categorical properties define the nature of an object, while dispositional properties describe an object's potential to exhibit certain behaviors under specific conditions. For instance, "fragility" is a dispositional property that describes an object's potential to break under suitable circumstances. Unlike categorical properties, dispositional properties can change over time. However, it remains uncertain whether these approaches, by shifting the nomological power from laws to properties, make genuine progress. As mentioned in Chapter 2, some critics find property-talk mysterious and suggest thinking of the wave function in structural terms (French and Ladyman 2011). They view the wave function as part of the complex structure of the theory. Given the problems with structuralist accounts, I will not revisit them here.

The Wave Function as Knowledge

Another possibility is to think of the wave function as epistemic, as seen in IT approaches, Qbism, and relational quantum mechanics, which we briefly mentioned in Chapter 2. The main idea behind these interpretations is that *the wave function represents our best knowledge of the system under investigation*, not something objective in the system. By appearing in the Born rule, the wave function determines our expectations about measurement outcomes based on what we already know.

These approaches have faced numerous criticisms, most notably the inability to account for *interference phenomena*. If the system is in a definite but unknown state—that is, if the wave function is in a superposition because we are unsure of the system's state, rather than because the system possesses truly indefinite properties—why do we observe interference patterns? In other words, interference arises from the terms of the superposition being actual. For example, consider a water wave passing through two slits. The amplitude of the oscillations after the slits results from the combination of the amplitudes of the waves coming through each slit: they add up if they are in phase, and they subtract if they are out of phase. This is the essence of interference. However, if the wave function is epistemic, a superposition like $\psi_1 + \psi_2$ does not represent the analogous situation to the water wave. Instead, it reflects the fact that the wave function could either be ψ_1 or ψ_2, but we do not know which. There is no wave traveling through the slits, but interference still occurs. The question then becomes: where is this interference coming from? How could our ignorance about the true state of the system generate a physically observable phenomenon?

Moreover, epistemic approaches arguably *violate the PBR theorem*, formulated by Pusey, Barrett, and Rudolph (2012). This theorem has been interpreted to show that epistemic interpretations conflict with the statistical predictions of quantum mechanics. An epistemic wave function would necessitate the existence of impossible relations. Therefore, this theorem is often considered a no-go for epistemic views: the wave function must be ontic, meaning that it must objectively describe something—either a fact about matter, a nomological fact, or the like.

Aside from these issues, many realists are suspicious of epistemic approaches because they seem *too close to instrumentalism*, given their emphasis on measurement outcomes. However, as we noted earlier, some of these interpretations, particularly the IT approach, defend themselves as genuinely realist proposals. In their view, quantum mechanics is fundamentally concerned with measurement outcomes, and their statistics are recovered through the constraints that the Hilbert space structure imposes on quantum dynamics. These approaches aim to explain phenomena by systematizing them using general principles. As we saw in Chapter 2, this contrasts with the constructive understanding that I favor. In a constructive approach, macroscopic objects are composed of microscopic entities, and their behavior is dynamically recovered. Consequently, the constructive approach provides reasons why a phenomenon happens, whereas the alternative provides reasons for why we observe it.

The Wave Function as Prescribing Expectations

The prominent pragmatist quantum proposal is pragmatist quantum realism (Healey 2012), according to which the wave function prescribes what we should expect in a given physical situation. In this perspective, *the wave function is prescriptive of expectations* rather than descriptive of facts. This view is pragmatist in that the wave function is understood in terms of its roles in our scientific practice, specifically the function it plays within the theory. The wave function provides objective means to concisely summarize what happens and enables us to make predictions about what will happen next. Thus, the wave function's role is to explain why experimenters are justified in believing that something will or will not happen. Pragmatist quantum realism interprets quantum theory as a *principle theory*: the Born rule assigns probabilities to phenomena, and these probabilities express our degrees of belief that a given phenomenon will occur. Thus, pragmatist quantum realism is similar to epistemic views in terms of the type of explanation it provides. However, pragmatists do not prioritize recovering empirical data, nor do they assign a fundamental role to measurement results as epistemic approaches do. (For this reason, epistemic views might be considered pragmatic, rather than pragmatist.) Furthermore, pragmatists offer a characterization of the wave function that is not epistemic. Since they define the wave function in terms of its role in the theory, rather than as our knowledge of the system, the wave function is objective. Therefore, arguably, the PBR theorem does not apply to pragmatist quantum realism.

Thus, this approach shares with my own focus on explanation and the understanding of the wave function as not representing matter. However, the two views differ in what they consider a satisfactory explanation: while I favor a constructive approach, pragmatists, like epistemicists, do not believe constructive explanation is necessary and see principle theories as sufficient. As we have seen, principle theories explain phenomena by constraining them with principles, much like thermodynamics explains heat phenomena. A principle such as "heat flows from high temperatures to low temperatures" explains why my coffee cup gets cold when left on a table for a while—it is because heat flows from the hot cup to the cooler environment. However, as I have repeated multiple times, I think that constructive theories are more explanatory. For instance, I need to explain what heat is and what causes it to flow in that particular way. Otherwise, I would only have *a description*, not a deeper understanding of the phenomenon's origin.

My goal is to *reconcile the pragmatist quantum realist commitment to a nonmaterial, nonrepresentational, ontic wave function with the constructive explanation offered by the nomological approach.* In the next section, I do this by drawing on another philosophical position: functionalism.

Wave-Functionalism

The central challenge is, therefore, to develop an *ontic understanding* of the wave function that allows for a *constructive explanation without representing anything.* The nomological approach provides the correct explanatory framework, but in it, the wave function is still representational. Pragmatist quantum realism, on the other hand, supports a nonrepresentational stance but lacks a robust explanatory framework. My proposal offers a hybrid account that integrates key elements from both approaches, aiming to preserve their core motivations while addressing their respective shortcomings. I call this view *wave-functionalism*, as it defines the wave function in terms of the roles it plays within the theory: the wave function is whatever performs those roles. In what follows (see also Allori 2021b), I develop this approach in detail.

The Wave Function Is as the Wave Function Does

Broadly speaking, functionalism is the view that certain entities can be defined in terms of their functional roles. A powerful slogan to characterize this view is as follows: "The table is as the table does." If the function of the table is to hold stuff off the ground, then whatever performs that function is a table. Note that anything defined in functional terms is typically *multiply realizable*: a table, for example, can be square or round, have four legs or just one, be large or small, tall or short, and be made of wood, metal, plastic, or any number of other materials. Functionalism has a long tradition in the philosophy of mind. According to functionalist approaches, mental states are defined in terms of their functional roles, either reducing them to their functions or thinking of them as nonfundamental states, functionally emergent from physical states. Given their functional characterization, mental states can be realized by different physical systems. For instance, mental states are realized in humans by brain states, while they could be realized by positronic circuits in Commander Data. Recently, functionalist approaches have also been introduced in the philosophy of physics. For instance, Wallace (2012) uses functionalist techniques in the many-worlds framework to explain the emergence of objects: they are useful patterns, defined in terms of the functions they play. Moreover, as we saw in Chapter 3, Albert (2015) functionally defends wave function realism, arguing that three-dimensional objects are functionalized in terms of their causal roles. It can also be argued that spacetime itself is amenable to a functionalist treatment. For example, in classical gravity, Knox (2013, 2014, 2018) contends that spacetime is functionally emergent, grounded in its role of defining inertial frames. Lam and Wüthrich (2018) argue for the functional emergence of spacetime in quantum gravity.

My view applies functionalism to the wave function, "reverting" the arguments of Albert and Wallace, who wish to functionalize objects in terms of the wave function, hence the name wave-functionalism. I argue that this approach to the nature of the wave function captures many of the advantages of rival approaches without their weaknesses. In my view, the wave function should not be understood in terms of what it represents, as pragmatist interpretations suggest. We should neither seek a material entity, as wave function realism does, nor a nomological fact, as proposed by nomological approaches. Nor should we interpret it as a reflection of incomplete knowledge, as in epistemic interpretations. Instead, I propose that *the wave function is an objective element of the theory, defined functionally by the roles it plays within that theory*. By adopting this functionalist perspective, we can capture the intuition that the wave function is law-like—similar to nomological views—while avoiding issues such as its time-dependence and nonuniqueness. At the same time, we can understand the wave function as nonmaterial—like in epistemic views—but still ontic, thus sidestepping no-go theorems. Finally, this framework allows us to treat the wave function as nonrepresentational, in line with pragmatist aims, without reducing it to a mere tool for predicting expectations.

Let me now add two remarks. First, as we will discuss later, what is functionalized is the ingredient in the law which the wave function happens to realize. That is, *the wave function is one of the possible realizers*; another realizer could be a wave function with a different phase, or a density matrix. Second, this type of functionalism is *nonreductive*: the information contained in the wave function cannot be reduced to the information contained in the fundamental ontology of matter (see next section).

The Functions of the Wave Function

So, what are the roles the wave function plays in the theory that we can use to functionalize it? In the following, I will discuss a series of different but interconnected roles that together contribute to the same goal: to *effectively reproduce the empirical data in an explanatory way*.

Reproducing the data is what the wave function ultimately does. It is the state of any physical system in the sense that it contains all the information about them, but in my approach, it does not represent them, as it is not a spatiotemporal object. Therefore, this information needs to be extracted from the wave function to make contact with experimental results. This is the wave function's *empirical role*: producing the empirical predictions of quantum theory. This is done by the Born rule, which projects the information in the wave function into spatiotemporal entities—namely, empirical data one can observe in the laboratory. In other words, the wave function is part of the pragmatic mechanism that connects the formalism with experimental results. In this way, the theory can be thought of as a *black box* that takes in the description of the initial conditions and the initial system and provides us with a description of where the system has evolved into, in terms of three-dimensional macroscopic pointer positions. The wave function is part of the internal mechanism that makes this process work. This process is fundamentally *pragmatic*: it contributes to generating probabilities through the Born rule, but the wave function's role is not unique. For example, one could use a different wave function that differs only by a global phase and still generate the same probability distributions. This is why the state of the system is often characterized in terms of rays in Hilbert space, an equivalence class of wave functions related by phase transformations—any member of the class leads to the same empirical predictions. Moreover, one could formulate quantum theory using a density matrix instead of a pure-state wave function and still recover the same observable outcomes. The flexibility extends further. In theories like the pilot-wave theory, the dynamics are defined using a Schrödinger-evolving wave function. In contrast, GRW theories rely on a wave function that undergoes spontaneous and random collapses. Yet, it is possible to construct alternative formulations of these theories that reverse these roles: a GRW-like theory with a noncollapsing wave function, or a Bohmian-like theory with a collapsing one, both yielding the same empirical predictions (Allori et al. 2008). This shows that the choice of wave function behavior—whether collapsing or evolving smoothly—is not dictated by empirical necessity but by pragmatic and super-empirical considerations, such as *simplicity, coherence with other parts of the theory, and explanatory power.* For these reasons, the standard formulation of quantum theory with a Schrödinger-evolving wave function is often preferred: it is the simplest and most effective way to represent the dynamics of the theory. Thus, all these formulations—despite their different mathematical structures—are *effectively equivalent from a practical and explanatory standpoint*. They make the same predictions and account for phenomena in the same way. This means that the use of the wave function in the theory is not indispensable: what is indispensable is the information the wave function provides about physical systems, but not the wave function itself. In this vein, the wave function's role is therefore best understood functionally: it does what is needed to generate predictions, regardless of whether it is taken to represent a physical entity, a law, or something else entirely.

The empirical role of the wave function alone is sufficient to account for the observed experimental outcomes: they are what the Born rule prescribes. In both epistemic and pragmatist approaches to quantum theory, which have a large empiricist component, this is all that is required—the wave function is used to generate statistical predictions about experimental outcomes and thereby connect the mathematical formalism of the theory with observable data. However, if one's aim is not merely to predict outcomes but to explain them—specifically in terms of a constructive account of why those outcomes occur—then this empirical role is not the only one the wave function, or whatever encodes all the information about the system, performs. A constructive theory, as I have defended in Chapter 4, demands that the wave function play additional roles beyond the merely predictive. One key role the wave function assumes in this context is *nomological*. Within my perspective, this role is explicit and central: *the wave function governs the evolution of the fundamental ontology of matter* (whatever that is, as long as it is in spacetime), effectively serving as part of the dynamical laws. That is, the wave function is not itself a material entity but a mathematical object that appears in the dynamical law which guides the motion or development of the material ontology.

Notice that even in epistemic perspectives, the wave function can be seen as playing a nomological role insofar as it is embedded in the principles that define how the theory functions. Similarly, in pragmatist quantum realism, the wave function acts as an "information bridge," summarizing the relevant conditions and expectations without necessarily representing physical reality. It still helps determine what kinds of outcomes are likely and in what contexts, thus contributing to the explanatory structure of the theory. In all these frameworks, the wave function plays a key *explanatory role*, albeit in different ways. In my approach, the explanation is *constructive*, since the wave function helps define the laws that govern the behavior of matter at the most fundamental level. This provides an explanation of macroscopic phenomena in terms of the detailed behavior of a microscopic ontology. By contrast, in epistemic and pragmatist approaches, the explanation is principle-based: the wave function contributes to explaining macroscopic phenomena by constraining possible outcomes and providing statistical expectations. Here, the emphasis is not on compositionality and dynamical explanation but on the structural features of the theory that account for observed regularities. Hence, as we saw, epistemic views and pragmatist quantum realism are principle theories.

Taken together, these empirical, nomological, and explanatory roles constitute a functional definition of the wave function: it is identified by the functions it performs within the theoretical framework. This resonates with the broader functionalist strategy introduced earlier in the chapter. According to this view, to understand what the wave function is, one need not specify what it represents in ontological terms: that is, it does not matter what the wave function represents. I agree with pragmatist quantum realism, which maintains that the wave function is not something one needs to reify. The question "What does the wave function represent?" is at best misleading, as it shifts the focus away from the real question: "*What does the wave function do for us in the theory*?" Within my view, it is sufficient to identify the wave function's contribution to the nomological, explanatory, and predictive structure of the theory. Depending on the framework, these contributions may be purely empirical, deeply constructive, or somewhere in between. Nonetheless, in all cases, the wave function earns its place in the theory through what it does.

Realizing the Functions

As discussed above, wave-functionalism captures many of the *core motivations* behind interpretations that reject a literal, representational reading of the wave function. In the following, I expand on the details of this view, while later I will demonstrate how wave-functionalism *avoids the major pitfalls* that challenge alternative interpretations.

At the heart of my approach is the idea that the wave function is not itself a law, but rather realizes an important element within a law-like structure. Depending on the metaphysical account of laws one adopts, laws may be seen as patterns that systematize phenomena, as prescriptions that determine how matter evolves, or as tools for guiding expectations about experimental results. Regardless of one's stance, physical laws, as we write them in physics textbooks, require various *ingredients*: mathematical entities and variables that serve different purposes. Some of these elements are constants of nature, like Planck's constant or the speed of light. Others are dynamical variables that describe physical systems, such as mass and charge.

These *properties describe the nature of matter*: they allow us to distinguish an electron, say, from a proton. They also specify *how particles will behave* in a given situation: an electron turns right and a proton turns left in a magnetic field because one has a negative charge and the other a positive charge (however, see next chapter). Another important group of variables consists of *interaction-defining quantities*, such as force fields or potentials. For instance, two charged particles feel one another's influence as specified by the electric potential, which describes how one is affected by the other's presence. *The wave function*, in this picture, falls into the latter category: it *mathematically encodes how matter behaves and interacts*, serving as one of the necessary ingredients of the law.

This suggests an imperfect analogy between the physical law and a *birthday cake.* The cake's purpose is to bring joy to the birthday person, just as a physical law's role is to explain and predict physical behavior. To accomplish this, both require distinct ingredients that contribute in specific ways: sugar for sweetness, flour and eggs for structure and cohesion. Similarly, laws of nature require constants, interaction terms, and variables that together make the law operational. The wave function, in this analogy, is like a key flavoring or structuring ingredient: it's not the whole cake (i.e., the entire law), but it's essential for the cake to have the right effect (see Figure 5.1).

Another analogy (also imperfect but suggestive in another respect) is with a *symphony orchestra*. In this analogy, a physical law is like a musical performance: the resulting symphony (i.e., the observable phenomena) depends on many coordinated elements. The musicians correspond to the fundamental ontology, such as particles. However, not all musicians play the same instruments, just as not all particles are identical (however, see next chapter): particles have intrinsic properties, like mass or charge, which define them as such—an electron or a proton, say. Thus, there are different families of particles behaving differently from one another, just as there are different instruments in the orchestra playing different music sheets. Electrons behave differently than protons in the same physical circumstances, and violins play

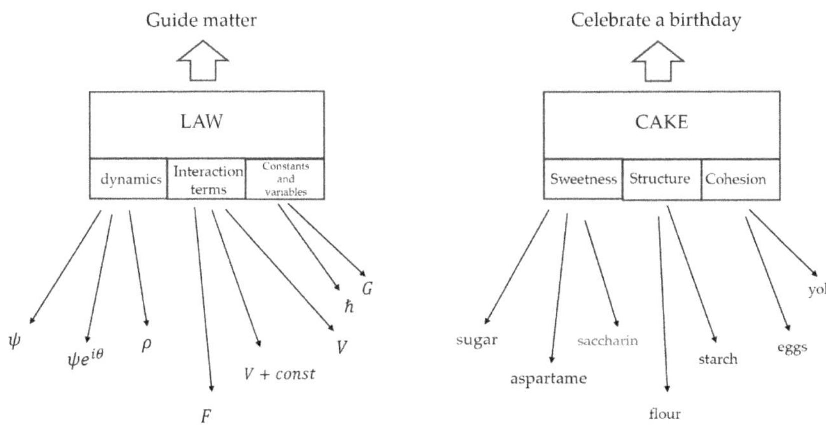

Figure 5.1 Multiple realizability of functions.

a different score than trumpets. The elements in the law that specify how the different families of particles behave are the ones that define the interactions. Similarly, in the orchestra, it is the music sheet played by a given instrument that differentiates one instrument from another. This analogy shows how the wave function is not a musician or an instrument. Rather, it is more like a music score: it does not produce sound on its own, yet it plays a central role in determining how the performance unfolds. Likewise, the wave function does not describe material entities, but instead helps orchestrate the evolution of the system by guiding the behavior of matter. Even in early formulations of the pilot-wave theory, the wave function contributed *via* the quantum potential, which influenced particles as if it were a force-like field (Bohm 1952). Later formulations abandoned the quantum potential concept due to difficulties in interpreting its physical basis (e.g., no associated charge or source). Yet, the wave function remains crucial in the guiding equation, showing that it still plays a nomological role: it governs motion, even if not as a force in the traditional sense. Thus, wave-functionalism proposes that the wave function is best understood not as a material entity, nor merely an expression of our knowledge, but as a theoretical component defined by the roles it plays in generating predictions, explaining outcomes, and structuring the theory. Its status is functional: just as a musical score ensures that the different instruments in the orchestra play harmoniously without making music directly, the wave function enables the physical system to evolve coherently, even if it doesn't itself belong to the "orchestra" of material ontology.

One might press for clarification on *what it truly means to say that the wave function is "a part of the law."* At first glance, this may seem obscure or even incoherent: laws of nature are typically conceived as unified entities, not as composites. Whether they are viewed as governing the behavior of matter, summarizing patterns in a Humean mosaic, or constraining possible events, laws are not usually thought of as having parts in the ordinary sense. The language of "ingredients" or "components" might thus appear misleading or metaphorically vague, raising the concern that this account simply replaces one puzzle (e.g., a time-dependent, nonunique, abstract nomological entity) with another (a "law" composed of unspecified "parts"). However, this concern can be addressed by focusing on how laws are actually expressed in physical theories. When physicists formulate laws in practice, they do so by using a collection of mathematical symbols and structures: constants (e.g., $\hbar, G$), parameters (mass, charge), fields (like the electromagnetic field), potentials, and functions such as the wave function. These are not arbitrary symbols, but essential components of the mathematical machinery that allow the laws to do their work, whether it is predicting outcomes, prescribing dynamics, or summarizing empirical regularities. As emphasized above, the ingredient realized by the wave function could also be realized by other mathematical objects. Therefore, in a nutshell, to say that the wave function is "part of the law" is to say that it is one of the possible realizers of certain functions that appear in the law; it performs a specific functional role within the structure of the theory. And crucially, that role could in principle be played by other mathematical entities. That is, the function is multiply realizable. For instance, the Born rule can be formulated using a wave function, a density matrix, or even a wave function differing by a global phase. Since each of these objects serves the same function—namely, determining the probabilities of measurement outcomes—they are possible realizers of the same function. To understand this, consider the cake analogy again. Some ingredients in the cake may not be indispensable in the sense that their function may be multiply realized: white sugar can be substituted by cane sugar; flour with starch, butter with margarine, eggs with just yolk. In the orchestra analogy, one could substitute a violin with a cello, a trumpet with a tuba, or an oboe with a clarinet. Or, to use a different analogy altogether, consider a logical circuit. The output of the circuit depends on the roles played by various gates (AND, OR, NOT, etc.), but those roles can be realized by different physical implementations: silicon transistors, vacuum tubes, or even water valves in a hydraulic model. The same logical structure—the same "law"—can be implemented in different ways. Likewise, the laws of physics may include structures whose functions can be fulfilled by different mathematical entities. In this sense, what is being functionalized is not the wave function itself, but the ingredient in the law that the wave function happens to realize. Therefore, when we say the wave function is "part of the law," we mean it realizes a particular nomological function, such as encoding interaction dynamics or generating probability distributions. And just as the same logical gate can be built from different materials, the same nomological role might be fulfilled by other mathematical structures. What matters is not the object's intrinsic nature, but its role in the system that yields empirical predictions. This shift from substance to structure, from representation to function, is what wave-functionalism highlights.

That said, some might contend that whenever a law is formulated using different realizers or components, we are in fact dealing with a different law and, consequently, a different theory. Take, for instance, the pilot-wave theory: when expressed through the guidance equation, it appears distinct from the version articulated via the quantum potential. The two formulations involve different ingredients, and thus, one might be tempted to conclude that they represent two separate theories. But this conclusion seems misguided. In response, one could argue that these are not distinct theories because, at bottom, they do not posit different laws. What we have here are two alternative representations of the same law, each articulated in terms of different realizers. Importantly, even when the law is expressed using different components or mathematical structures, its substantive content remains unaltered. That is, the law retains its identity so long as it yields the same behavior of matter, organizes empirical phenomena in the same systematic way, or serves the same normative function in guiding agents' beliefs. As emphasized earlier, *a law is what it is independently of how it is written*. One can express Newton's equation using a certain potential, or that same potential with an added constant; yet these variations do not change the law's essence. Similarly, the nomological role of the wave function can be represented within the quantum potential framework or through another structure entirely. What matters is not that the wave function itself carries this role, but that the relevant content is realized in some suitable way. The identity of the law hinges on what is realized, not on the particular entity doing the realizing. That is, *the ingredients in the law are multiply realizable, because laws are understood functionally too: their role is to tell matter how to move.* This perspective should not seem especially foreign to a Humean. Within the Humean view, laws are not metaphysically primitive but are rather convenient summaries: they are patterns that, using the common God metaphor, God might print on a T-shirt to concisely describe the world's workings. Certain strands of Humeanism (such as Esfeld's super-Humeanism [2014a], Callender's account [2015], and the version developed by Bhogal and Perry [2017]) embrace this flexibility. They depart from the more traditional Lewisian framework by relaxing the requirement that fundamental laws refer solely to perfectly natural properties instantiated in the Humean mosaic. By contrast, Lewisian Humeans may reject this flexibility. On their view, the best systematization of the world might well include the Schrödinger equation as one of its axioms. Since the Schrödinger equation explicitly refers to the wave function, they would conclude that the wave function must be part of the Humean mosaic itself. In this way, the Lewisian approach seems to assign more ontological significance to the wave function, whereas the more permissive Humean views regard it as one possible representational tool among others.

To conclude, I would like to make a remark about *reductionism* in this context. Functionalist approaches are often reductionist concerning the entities they functionalize. That is, such approaches typically deny the existence of the functionalized entities as anything over and above their functional roles. For example, in the context of the philosophy of mind, functionalism holds that mental states are nothing more than certain configurations or processes of physical states; they are fully accounted for by the physical realizers of those functions. Similarly, in Albert's version of functionalism (Albert 2015), the three-dimensional macroscopic objects we ordinarily think of are not posited as independent ontological entities existing in addition to the wave function; rather, they emerge from the dynamics of the wave function itself. Wave-functionalism, in this spirit, treats the wave function not as a fundamental entity in its own right, but as something that is reduced to the set of roles it plays (nomological, dynamical, representational, or otherwise). However, and this is crucial, wave-functionalism does not embrace reductionism in every respect. Specifically, *it does not reduce the realizers themselves*—namely the wave function and other elements featured in the laws—to something else, such as the distribution of matter alone. These components must still be posited in order for the laws to function in the way they are meant to. In other words, the realization of the functions attributed to the wave function and related entities requires their existence in some form beyond just the material configuration of particles. This is analogous to saying that God, when creating the universe at the beginning of time, did not merely set up a distribution of matter but also established the laws governing that matter. Those laws, in turn, contain the requisite ingredients, such as the wave function or something that fulfills its role, that enable them to guide or structure the behavior of the world in the way they do, regardless of what one takes that nomological function to be. The wave function offers one particularly effective and convenient way of implementing these functions, but as we have seen throughout, it is by no means the only possible way. Other formulations might serve the same purpose, and from a functionalist perspective, what matters is the realization of the roles, not the specific entities that fulfill them.

Philosophical Payoffs

I now turn to consider how wave-functionalism offers solutions to several of the difficulties faced by alternative interpretations concerning the nature of the wave function. I begin with the strong nomological approach. In that framework, the wave function is taken to represent the law itself, thereby granting it a uniquely privileged status. In contrast, wave-functionalism treats the wave function as one of potentially many possible realizers of a certain component of the law. In this view, *the wave function does not need to be unique*, just as potentials in classical physics are not unique. A potential is properly understood as an equivalence class of entities that differ by a constant; analogously, the wave function constitutes an equivalence class of entities that differ by a global phase. Since wave functions that differ only by a phase yield identical observable outcomes, they are equivalent for all explanatory and predictive purposes. Furthermore, wave-functionalism is not troubled by whether or not the wave function evolves over time. This question becomes problematic only if one assumes that the wave function is itself a law. But within the wave-functionalist framework, the wave function is simply a component in the mechanism that gives rise to observable phenomena and underlies our explanatory models. Just as certain scenarios in classical physics require *time-dependent potentials*, in quantum mechanics we may require a time-evolving wave function in some contexts. The essential point here is that the wave function fulfills a nomological role, as it contributes to the workings of the law. Nonetheless, it is not itself a law. This distinction allows wave-functionalism to sidestep many of the difficulties encountered by the strong nomological interpretation.

In relation to the weak nomological approach, according to which the wave function is a property of matter, wave-functionalism distinguishes itself by avoiding any dependence on a theory of properties. *It does not require that the wave function be classified within any existing metaphysical category of properties*. In fact, as I will argue in the following chapter, it is possible to dispense entirely with property-talk and instead frame the discussion purely in terms of laws. If that is indeed the case, then there seems to be little justification for treating the wave function as a kind of property at all.

Wave-functionalism also shares with the epistemic interpretation the insight that the wave function does not represent material reality directly: it does not describe matter itself. However, unlike the epistemic view, my position does not treat the wave function as a representation of incomplete knowledge or belief about a system. Rather, the wave function is functionally characterized by the roles it plays within the theoretical framework. Principally, it is characterized by its role in enabling efficient and coherent recovery of empirical data. Because of this functional rather than epistemic description, on this view *the wave function is objective*: it exists independently of observers and their cognitive states. It is thus ontic without being material. This allows wave-functionalism to remain immune to challenges like the PBR theorem, which targets epistemic interpretations of the wave function.

Moreover, wave-functionalism overcomes a significant limitation faced by pragmatist forms of quantum realism, namely, the charge of being insufficiently explanatory. As discussed earlier, pragmatist and epistemic views can be suitably characterized as principle theories: they account for why experimenters are rationally justified in expecting certain outcomes, but they do not provide deeper explanations of why those outcomes occur in the first place. For those who seek to understand what causes a given phenomenon, what underlies or grounds it, or what mechanisms bring it about—namely, those looking for a constructive explanation—such accounts may ultimately prove unsatisfying. Wave-functionalism, however, is not bound by the limits of principle-based explanation. *It is flexible enough to support both principle and constructive explanatory models*. In this regard, it aligns with the explanatory goals of my preferred approach, as defended in Chapter 4. The wave function, in this framework, can be understood as realizing part of the dynamical law that governs the motion of microscopic entities and accounts for macroscopic behavior. To illustrate, consider the well-known interference and diffraction pattern observed in the two-slit experiment. One way to explain the resulting distribution of dots on the detection screen is to state that they conform to the probability pattern given by the squared modulus of the wave function. This explanation is compatible with wave-functionalism and exemplifies a principle-type account: the wave function plays a constraining role, delimiting the range of expected phenomena and telling us where the dots are likely to appear. However, *within wave-functionalism, a further, constructive explanation is also available*. One can say that the reason the dots appear where they do is because the wave function realizes an ingredient of the law governing the actual motion of the particles. In other words, wave-functionalism can accommodate both kinds of explanation: it allows for predictive, principle-based reasoning, while also making room for deeper, causal, or dynamical accounts. This sets it apart from pragmatist quantum realism, which remains restricted to the former. Wave-functionalism's greater flexibility in explanatory strategy thus constitutes a key advantage over competing interpretations.

Potential Criticisms and Clarifications

In this section, I will address *several objections* to wave-functionalism that have not yet been discussed. Some of these concerns arise from misunderstandings of my position, and the responses I provide will also help clarify the view I am proposing.

One concern that might be raised is that wave-functionalism may not be able to solve the *configuration space problem*. We discussed this problem in Chapter 3 when we noted that the wave function, residing in a high-dimensional space, cannot be straightforwardly viewed as a three-dimensional wave. My spatiotemporal foundational ontology approach is meant to solve this by stipulating that matter is directly in spacetime. The worry here is that wave-functionalism might reintroduce this problem: after all, nomological facts are said to exist in this space. To better understand this objection, consider the double-slit experiment. A wave-function realist can offer an explanation: there is a physical wave that passes through both slits, generating two waves that interfere and produce the diffraction pattern on the screen. On the other hand, the epistemicist faces more difficulties: if the wave function represents incomplete knowledge about the system, how can it generate a physical phenomenon? Wave-functionalism could explain the interference pattern by treating the wave function as a potential, though this would require defining it over configuration space. This raises the concern that, in doing so, wave-functionalism might be privileging configuration space as the appropriate domain for the law. Since one of the motivations behind wave-functionalism, and my entire approach, is to avoid relying on configuration space, this could be seen as a problem (Lewis, p. c.). In response, I acknowledge that, given that the primary innovation of quantum mechanics is nonlocality, some reference to configuration space is *inevitable* when describing quantum systems. Everyone, including me, must address this issue in some way. However, wave-functionalism does not claim that there is a physical ontology in configuration space. Rather, it asserts that nonlocal nomological facts are encoded in, or realized by, the wave function. This seems far less problematic than claiming that physical objects themselves exist in configuration space. In fact, my perception of material bodies occurs much more directly in three-dimensional space than my perception of laws, which I observe only indirectly through the effects of those bodies. I believe it is more justified to say that material bodies interact through a law that involves an ingredient located in configuration space, rather than claiming that those bodies themselves exist in that space. Laws, after all, are not meant to be physically located in space. From a Humean perspective, it is also easier to conceptualize a law as existing in configuration space rather than physical objects. Indeed, a law that operates in configuration space is one where the interactions it governs are more complex and interconnected than one might have initially imagined. The various approaches taken by wave-function realists have shown that it is much more challenging to explain what it means for a physical object to "vibrate" in configuration space. Given that every interpretation of quantum mechanics must somehow address the issue of configuration space, wave-functionalism appears to offer the best solution by incorporating it within the law itself, rather than within the material ontology. This approach aligns well with the nonlocality inherent in quantum theory.

Another potential objection is that wave-functionalism *fails to provide an adequate explanation* because the wave function does not materially exist. Historically, in science, new entities have been introduced to account for unexplained phenomena, such as neutrinos, electromagnetic fields, and dark matter. It could be argued that the wave function was similarly introduced in quantum theory to explain phenomena like the double-slit experiment. However, critics may ask: if wave-functionalism claims that the wave function does not correspond to any material reality, how can it possibly explain the observed phenomena (Maudlin, p.c.)? To address this concern, I would point out that pragmatist quantum realists face the same challenge. Defenders of these views argue that their interpretation does not explain phenomena by describing the physical world, but by providing reasons to expect certain experimental outcomes. I do not believe it follows that every component of an explanation must be material. For example, electromagnetic fields were initially introduced as useful mathematical devices, without any assumption that they represented material fields. One could have raised the same objection at the time, but no one questioned whether classical electrodynamics was explanatory. The reason, I believe, is that even if electromagnetic fields were understood as mere fictions, they still contained crucial nomological information necessary for the theory to generate empirically adequate predictions. In other words, the realism in classical electrodynamics was about the laws, specifically the nomological facts, rather than about the specific entities that appeared in those laws. Wave-functionalism follows a similar logic: *it is the law itself, rather than the wave function specifically, that is explanatory.*

Key Takeaways

In my approach, wave-functionalism, I propose that the wave function should be understood in terms of the functions it serves within the theoretical framework. Rather than being viewed representationally, the wave function is defined by the various roles it plays within the theory. These roles are diverse yet interlinked, and they include an empirical role, where the wave function helps generate experimental predictions; a nomological role, where it appears in the evolution equation of matter; and an explanatory role, where it contributes to constructing explanations for physical phenomena.

Within this framework, I argue that the wave function, when understood functionally, possesses several distinctive characteristics. First, it is *ontic without being material*, in contrast to wave-function realism, which treats the wave function as a material entity in the world. Second, it is *nonmaterial without being representational*, unlike pragmatist quantum realism, which views the wave function as a tool for representing our knowledge of a system. Additionally, the wave function is *nomological without being a law*, distinguishing it from the traditional approach of primitive ontology theories, which treat the wave function as a law in itself. Finally, it provides a constructive, mechanistic explanation of phenomena, aligning with the goals of the spatiotemporal foundational ontology approach.

By maintaining these features, wave-functionalism is able to *sidestep major objections* faced by competing theories. For example, by having a nomological role without being a law, concerns about the wave function's time evolution or uniqueness become irrelevant. Since it is not a law but rather part of the mechanism that governs physical behavior, these issues are not problematic. Additionally, its explanatory role is linked to laws of nature, so the wave function is not confined to the abstract explanations of principle theories. Its explanatory power extends beyond mere prediction, offering a more robust and constructive explanation of the phenomena observed. Moreover, wave-functionalism avoids the challenges posed by no-go theorems by remaining nonrepresentational and nonepistemic. Unlike wave function realism, which treats the wave function as representing material objects, or pragmatist quantum realism, which reduces it to incomplete knowledge, wave-functionalism situates the wave function as an objective component within the theory. This ensures that the wave function fulfills its function in explaining and predicting physical events without encountering the issues raised by other interpretations.

In conclusion, wave-functionalism presents a comprehensive and detailed account of the wave function that is both objective and explanatory. By defining the wave function in terms of the roles it plays within the theory, rather than treating it as a material entity or an epistemic tool, wave-functionalism provides a robust framework capable of addressing the concerns of those seeking a constructive and objective explanation of quantum phenomena, while avoiding the pitfalls of competing theories.

In the following section, I will examine which form of realism aligns with this perspective on the nature of the wave function, and explore how one might counter the pessimistic meta-induction argument against realism in the shift from classical to quantum theory, emphasizing continuities rather than revolutionary changes.

Restricting Realism in the Quantum Domain

As discussed in Chapter 1, scientific realism is challenged by the *pessimistic meta-induction* argument: contrary to the realist view, the success of a theory does not necessarily indicate its truth, as past successful theories have ultimately turned out to be false. A potential response to this challenge is to limit the scope of realism, as proposed by structuralism and *explanationism*, among others. According to explanationism, which was developed outside the quantum domain by Psillos and Kitcher, we are justified in being realists about the entities responsible for a theory's success if these entities are preserved in theory change (see Chapter 1). The core idea is that only certain parts of the theory carry out the explanatory work, and these parts are responsible for the empirical success of the theory. These essential components are the *working posits* of the theory—they are the pillars that make the theory function. Other aspects of the theory, however, are merely provisional assumptions, which explanationists call *presuppositional posits*. These elements are necessary to complete the theory after determining the ontology, and they fill in the gaps required to account for observed phenomena. However, they do not have ontological significance. Explanationists argue that these "idle wheels" do not need to be retained during theory change in order to justify realism. As long as the working posits remain in the new theory, the pessimistic meta-induction argument can be successfully countered.

Without delving too deeply into the specifics of the original proposal, I will reframe this idea within the framework I am defending, as I do in this section. I argue that, within my spatiotemporal foundational approach, *the wave function is best understood as a presuppositional posit*. The fundamental ontology of matter, however, is the working posit of the theory: it is through its dynamics that we can explain macroscopic phenomena compositionally and dynamically. If we adopt wave-functionalism, it is clear that the wave function plays a role in realizing the laws of nature. While laws are essential for explaining the behavior of matter, the specific realization of these laws is not of primary concern. Since the wave function merely serves as a realizer, it can be seen as a presuppositional posit—something provisional that is likely to be replaced by the next best theory, and something we do not need to be ontologically committed to.

Therefore, the pessimistic meta-induction argument can be overcome if we can show that the fundamental ontology of matter, rather than the wave function, is preserved during theory change.

The Wave Function as a Presuppositional Posit

Let us explore in more detail how the fundamental material ontology of quantum theory, as proposed in Chapter 4, can be identified with the working posit of quantum mechanics, while the wave function is best viewed as a presuppositional posit. The *predictions of the theory are determined by the entity that describes the fundamental material ontology*, not by the wave function. In fact, the data are encoded in the former, not the latter: the pointers that indicate results are made of matter, not wave functions. Once the fundamental constituents of matter and their law of evolution have been specified, everything else follows. This includes making future empirical predictions and explaining past results.

The wave function appears in the derivation of the theory's predictions, but the way it specifies the law of matter remains underdetermined (Allori et al. 2014). As we've seen, one could predict and explain the same data using a different wave function shifted by a phase, or a density matrix, or some other mathematical tool. Furthermore, the evolution equation that the wave function obeys is also underdetermined by the data. Arguably, the Schrödinger equation was chosen for convenience. In fact, Schrödinger initially proposed a more complicated equation but later revised his approach (1926). This *underdetermination* is not problematic: as we have seen, a theory of particles following certain trajectories, such as the pilot-wave theory, can be formulated with a Schrödinger-evolving wave function, as in the usual formulation, or with a collapsed wave function (Allori et al. 2008). Therefore, the specific variables used to specify the law of evolution of the fundamental ontology of matter and how it evolves are not essential, as long as the governing law remains unchanged. More generally, in my framework, two theories that predict the same behavior for the fundamental ontology of matter—regardless of how this behavior is derived (e.g., via a Schrödinger-evolving wave function or not)—describe the same physics and are physically equivalent. For all explanatory purposes, they are the same theory. Since different wave functions can produce the same behavior for the fundamental ontology of matter, and since it is this behavior that accounts for macroscopic regularities, the wave function is not a working posit and should not be taken "too ontologically seriously." In fact, it is possible to attempt to eliminate the wave function entirely, as some have tried to do (Dowker and Herbauts 2005, Norsen 2010a).

Moreover, the explanatory power of the theory is constructive, meaning it is rooted in the fundamental ontology of matter, not the wave function itself. As we discussed in Chapter 1, in classical mechanics, macroscopic objects are understood as collections of particles, and their properties are explained by the interactions between these particles and their environment. For example, the color of a leaf can be explained by the way light interacts with the particles that compose the leaf, where certain wavelengths are absorbed and others reflected, giving the leaf its green color.

Similarly, the fundamental ontology of matter within my spatiotemporal framework serves as the foundation for the explanatory structure of quantum theories. In this framework, the properties of macroscopic objects are, in principle, explained by the behavior of their microscopic constituents. *The wave function plays a role in this explanation, but it is just one element within the broader set of laws* that govern the system, rather than being the primary or sole explanatory factor (Allori 2013a, b; see also Chapter 4).

As mentioned earlier, the distinction I have made between matter and the wave function aligns well with the idea that working posits are the entities that play a fundamental role in the theory's explanatory schema, while presuppositional posits do not. However, in contrast to Psillos and Kitcher's proposal, my approach explicitly postulates *a priori* what in the theory represents matter and what does not—such as the case of the wave function. I maintain that one should be a realist about the entities that describe the fundamental ontology of matter, regardless of what one believes the wave function really is: a law-like object, a property, or a new kind of entity. In the first part of this chapter, I argued that the wave function should be characterized functionally and that it does not represent anything material. This is consistent with the conclusions drawn in this section. First, the wave function, being one of the possible realizers of the functions it plays in the theory, is underdetermined by the data in an unproblematic way. As emphasized, one could be *metaphysically neutral* with respect to the wave function—one does not have to commit to it in order to account for the success of the theory. To interpret experimental predictions, a clear understanding of matter and its interactions is necessary. In the quantum domain, one could use the wave function or a density matrix to describe interactions, just as one might use forces or potentials in classical mechanics. In my spatiotemporal approach based on wave-functionalism, the realizers of the interaction are not entities that should be reified into something we encounter in the world. Therefore, the wave function is best viewed as a presuppositional posit: the working posits, which ground the predictions and explanations of the theory, are given by the variables defining matter. For these reasons, I believe that my spatiotemporal foundational approach provides an effective framework to extend explanationism into the quantum domain, thus producing a robust form of realism based on constructive explanation. This framework aligns with past explanatory models and respects symmetries.

Classical-to-Quantum Ontological Continuity

These reflections can be used to formulate a response to the pessimistic meta-induction argument, particularly if one can demonstrate that the fundamental ontology of matter remains intact through episodes of theory change. Let us consider the transition from classical mechanics to quantum theory. It can be argued that, within classical mechanics, matter is fundamentally composed of particles—entities that possess the intrinsic property of occupying definite positions in three-dimensional space. In the pilot-wave theory, this same conception of matter as composed of particles is preserved: particles continue to exist and retain their positions in physical space, guided by the wave function. Thus, one can maintain that there is *an ontological continuity* from classical to quantum theory, at least under this interpretation. This continuity provides a potential way to resist the pessimistic meta-induction, which highlights the history of abandoned theoretical entities to undermine scientific realism (see Chapter 1). In this context, by showing that the ontological core—namely, matter as composed of particles—is not discarded but rather carried forward into the quantum framework, one can argue for a stable realist commitment across theory changes. This, in turn, could be taken as an argument against quantum theories that do not include particles as part of their ontology, further supporting particle-based interpretations like the pilot-wave theory (see Figure 5.2).

However, one might contend that the key factor in mounting an effective response to the pessimistic meta-induction is not necessarily the preservation of a particle ontology, but rather the *retention of a spatiotemporal ontology*, as long as it suitably reproduces a particle ontology at the relevant scale. According to this view, even if a quantum theory posits a matter field ontology or a flash ontology, and thus does not carry over the particle framework from classical mechanics, it may still fare better than wave function realism in addressing the pessimistic meta-induction. In other words, one might argue that the shift from classical particles to quantum nonparticles does not represent a radical rupture, so long as the theory maintains entities located in ordinary three-dimensional space. In this view, even though quantum mechanics might ultimately reveal that there are not fundamentally particles, it still posits the existence of something that is *effectively localized* in three-dimensional space—some sort of distributed material presence or discrete events that, in certain respects, mimic the behavior traditionally attributed to particles. Thus, while our classical conception of matter as made up of discrete particles might be mistaken, the deeper assumption that reality is composed of entities situated in familiar spatiotemporal structures remains intact. This continuity—anchored in the three-dimensional nature of what exists, even if not in its particulate character—can be seen as sufficient to resist the pessimistic conclusion that all theoretical entities are likely to be discarded.

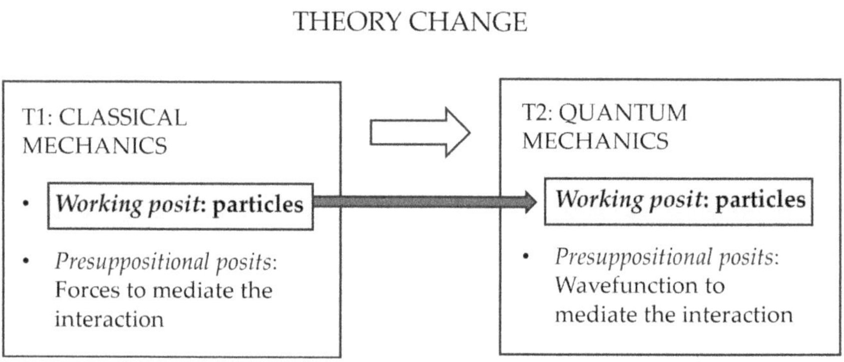

Figure 5.2 Working posits are preserved in theory change.

Let me expand on this point further. Consider a historical example from the development of atomic theory: scientists initially believed that atoms were the smallest, indivisible constituents of matter. Later, it was discovered that atoms themselves are composed of smaller, indivisible particles such as electrons, protons, and neutrons. In this case, the earlier theory was not entirely mistaken: it correctly posited the existence of particles, but it was incorrect about the precise identity and structure of those particles. What changed was not the ontological category (particles), but our understanding of the specific particles that constitute the fundamental building blocks of nature. Now, contrast that situation with a shift from a theory based on particles to one in which the fundamental entities are conceived differently altogether. This would be more akin to a transition from a quantum theory grounded in particle ontology to a theory like string theory, where the basic elements are no longer zero-dimensional particles but instead one-dimensional vibrating loops or perhaps even higher-dimensional extended objects. Here, the departure is deeper: we are not merely refining our understanding of what counts as a fundamental particle, but are adopting a completely different framework for describing the basic ontology of the universe. Still, one might argue that even in such a transition, there is a kind of continuity if we *consider how these new entities behave at certain scales.* This is the essence of what Schrödinger suggested when he proposed that particles are not truly fundamental entities, but rather emerge as localized wave packets. When viewed from a sufficiently distant or coarse-grained perspective, the intricate internal structure of these wave packets becomes invisible, and they appear as though they were simple point-like particles. In this sense, they can be treated as point-particles for explanatory purposes, particularly when accounting for macroscopic regularities and phenomena. The finer details of what these entities truly are can be ignored at that explanatory scale, just as one might, from a distance, mistake a garden hose for a one-dimensional line rather than a two-dimensional tube, or mistake a subway poster for a continuous image when it is, in fact, composed of discrete colored dots. Similarly, whether the fundamental quantum ontology is described in terms of discrete flashes, a continuous matter field, or some other entity entirely, there is a mesoscopic scale at which these elements give rise to regular patterns of behavior that are best interpreted as trajectories of particles. This emergent behavior allows us to retain a constructive explanatory framework even if, strictly speaking, the fundamental microscopic ontology is not particle-based. The key point is that, *as long as the underlying entities exist in spacetime and exhibit effectively localized behavior, they can still ground a particle-like explanation at higher scales.* Thus, from the mesoscopic level upward—including the macroscopic domain of ordinary experience—explanations can proceed as though particles were still the operative constituents of matter, even if, at the most fundamental level, this is no longer the case.

It is important to note that this line of reasoning is not available to proponents of wave function realism. One might initially think that wave function realists could adopt a strategy similar to the one I have outlined by arguing something like the following: "In classical mechanics, we mistakenly took the fundamental entities to be particles, but later we realized that, in fact, these entities are patterns within the wave function.

Nonetheless, we are not completely off the mark: there is still a scale at which these patterns behave as though they were particles, thereby preserving a degree of continuity." However, this analogy does not hold upon closer scrutiny. The key distinction lies in the absence of a fundamental microscopic spatiotemporal ontology. In the argument I presented, continuity is secured precisely because the fundamental ontology remains embedded in the same spatiotemporal framework that grounds macroscopic phenomena. The entities postulated at the fundamental level, whether particles, fields, or flashes, exist within the same three-dimensional (or four-dimensional) space that our everyday experiences are situated in. This shared spatial framework allows for a seamless transition between scales: by "squinting," or abstracting away from microscopic detail, we can recover the familiar macroscopic world as a coarse-grained manifestation of the underlying ontology. Wave function realism, on the other hand, does not offer this same luxury. As discussed in Chapter 3, the fundamental ontology posited by wave function realists resides in configuration space—a high-dimensional space that bears no direct resemblance to the three-dimensional world we perceive. Consequently, the move from the wave function in configuration space to the macroscopic structures we observe is not a matter of simply looking from afar and interpreting fine-grained structures as point-particles. Instead, it requires a complex, nontrivial, principle-like reconstruction of three-dimensional appearances from an ontology that is not itself embedded in three-dimensional space. That is, as discussed in Chapter 3, wave function realism must provide principles to bridge the gap from high-dimensional space to three-dimensional space before constructive explanations can be applied, but these very principles remain unexplained. This undermines the idea that wave function realism can appeal to the same kind of ontological continuity across theory change, since the spatial disconnect between its fundamental ontology and the observable world is precisely what prevents a straightforward emergence of macroscopic structures.

Potential Challenges and Responses

Having explained how explanationism works in the quantum domain, let us turn to *potential difficulties.* A general concern arises from the criticism that the distinction between working and presuppositional posits, as drawn by the explanationist strategy, is essentially retrospective. According to this criticism, we can only determine what constitutes a working posit once we know that a theory has been replaced. In other words, *working posits are retrospectively identified* as those retained by successor theories, while presuppositional ones are those discarded. This raises a serious concern: if the theory has not yet been superseded, we lack any principled way to know which elements we are justified in being realist about. This undermines the explanationist strategy as a tool for guiding present realist commitments (see, for instance, Stanford 2003). However, this concern does not apply within my framework, in which *the fundamental ontology of matter is explicitly posited* when the theory is constructed. It is not extracted from the formalism after the fact, nor is it inferred only upon theory replacement. Rather, the fundamental ontology is proposed upfront as that which best satisfies a combination of empirical adequacy, simplicity, explanatory strength, and symmetry. In this way, the distinction between working and presuppositional posits is drawn at the point of theory proposal, enabling a realist to identify from the outset which aspects of the theory they are committed to as reflecting reality, and which are better understood as instrumental or heuristic.

Another consideration worth addressing concerns the suspicion that my proposal might amount to *a form of ontic structuralism*. If we are willing to abandon commitment to the specific nature of fundamental entities across theory change, does this not imply that what we preserve is merely the structure, rather than any substantial ontology? To clarify, if structuralism is understood as the eliminativist view that there are only structures and no individuals occupying them, then my proposal clearly does not fall into that camp. My approach begins with the assertion that there are concrete physical entities, and that these form the basis of any acceptable physical theory. Nevertheless, what about the more moderate version of ontic structuralism discussed in Chapter 2? This view allows for the existence of individuals but maintains that they have no intrinsic identity and are instead defined entirely by the relations they bear to others—relations given, in the quantum case, by the wave function. Indeed, Esfeld (2014b; see also Esfel and Lam 2008, 2011) proposes precisely this: that in *moderate ontic structuralism*, the *relata* of quantum structure are the elements of the fundamental ontology of matter. On his account, we are realists about both the fundamental material entities and the relational structure among them, encoded in the wave function. Instead, my approach takes a different route. While there may be a structuralist dimension to wave-functionalism, as I will explore more deeply in the next chapter, its strength, especially in addressing the pessimistic meta-induction argument, resides in its treatment of the wave function as a presuppositional posit rather than a working one. This is crucial. Only by recognizing the wave function as a presuppositional posit can we establish ontological continuity across theory change. In contrast, Esfeld's account seems to imply that the wave function is part of the structure we must be realist about. But if this is the case, it becomes hard to explain where that structure comes from when considering classical physics, where the wave function does not appear. Since Esfeld (2014a) has previously argued for a Humean reading of the wave function (as something merely describing regularities rather than a concrete entity), it seems better to interpret his structuralism as *architectural* rather than eliminative. On this reading, the fundamental ontology of matter lies at the foundation of the physical world, while the wave function contributes to the architecture that relates those elements and generates the observed macroscopic phenomena. Thus, the focus remains on matter, and the wave function need not be understood as a working posit. This reading allows us to defend against the pessimistic meta-induction in a way that Esfeld's version does not. A related perspective that bears resemblance both to my view and to structuralism is Myrvold's *principle of metaphysical continuity*, which we discussed earlier: namely, that one should adopt a metaphysical framework that is less vulnerable to disruption during theory change. The core idea is that committing to an ontology of substances is unnecessary, and that metaphysical attention should instead be directed toward the distribution of properties (2019). The insistence on substances, according to this view, makes metaphysics less resilient in the face of scientific progress. Myrvold's principle plays a valuable role in reinforcing realism, particularly by offering a way to resist the pessimistic meta-induction. It is minimalist in spirit, aiming to track a persistent thread that weaves through the historical development of physical theories. We are encouraged to be minimalist about a core metaphysical set, namely those elements that have endured across successive scientific frameworks. This stance aligns with structuralism in its move to demystify any underlying substratum or *relata*, instead emphasizing the significance of properties and relations. It also shares common ground with my own view in that it grounds metaphysics in spacetime, thus preserving a meaningful link with previous theories. However, the key difference lies in the ontological commitment. Myrvold's account *dispenses with substance ontology* altogether, remaining uncommitted about what it is, which in turn obscures what, precisely, the properties are properties of. This absence undermines explanatory clarity. While minimalism is a virtue, here it risks becoming too sparse to be able to provide a constructive picture. In contrast, my approach seeks a *more balanced position*: one that retains enough ontological structure to remain explanatory, while still achieving continuity and resisting overcommitment.

Let me now turn to another objection, one that challenges the substance of the continuity claim I've been defending. It might be said that I have made things too easy for myself by focusing on the transition from classical to quantum mechanics in the context of a particle ontology. Of course, one might say, if quantum theories of particles (like pilot-wave theory) are deliberately designed to preserve the classical particle ontology, then it is no surprise that they exhibit continuity. But *this makes the continuity trivial rather than revealing*. A more challenging and informative case would be to examine theories such as those involving phlogiston, caloric, or ether, and to give compelling reasons why these supposed entities are not to be regarded as genuine parts of the ontology of their respective theories. In response, let me emphasize that these cases have been discussed independently of quantum theory, as noted in Chapter 1. For instance, various philosophers (notably Psillos 1995) have argued that ether, phlogiston, and caloric should not be counted among the working posits of their theories. If these arguments are convincing, then it follows that those elements do not belong in the ontology either. In the case of the caloric theory, Psillos notes that the material interpretation of heat was not a working hypothesis but rather a theoretical gloss. The actual working posit was the concept of the quantity of heat, which need not be tied to any material substance. Furthermore, Psillos rejects the holistic theory of confirmation, which holds that all parts of a theory are equally confirmed by evidence. This allows him to maintain that even if the caloric theory was empirically successful and regarded as true, it need not imply that caloric existed as a material entity. A similar line of reasoning applies to phlogiston theory. Although it postulated the existence of an element released during combustion, it was ultimately replaced by the oxygen theory, in which combustion involves the absorption of oxygen rather than the release of any substance. Whether scientists at the time believed in the literal material existence of phlogiston is a matter for historians, but in any case, the general lesson is that what was central to these theories was the idea of an exchange process—something happening during combustion—not necessarily the existence of a specific material substance. The ether example is somewhat more complicated but leads to a similar conclusion. One can interpret classical electrodynamics in two different ways, each with different ontological commitments (as argued in Allori 2015a). On one interpretation, the theory is fundamentally about particles, and the electromagnetic field, including ether, is part of the nomological structure, not the ontology. In this view, ether and the fields are not working posits. Therefore, the bottom line is that it is reasonable to maintain that the ontology of all these theories is given by the classical particle, rather than ether, caloric, and phlogiston. These theories then evolved in light of empirical developments and gave rise to successor theories such as kinetic theory, oxygen theory, and relativity, which refine our understanding of the ontology while preserving key working posits.

Finally, one might raise the objection that the wave function should not be considered a presuppositional posit. According to Kitcher's account, working posits are those elements of a theory that are indispensable for explaining the phenomena. On this view, one might argue that *the wave function is a working posit* because it satisfies precisely that role: it is indispensable for making contact with experimental results, just as the particle ontology is. It allows us to account for why a particle takes one path rather than another in a magnetic field, and thus it plays an explanatory role on par with the matter ontology itself. Therefore, it should not be considered merely a presuppositional posit but rather a working one (Emery, p.c.). In response, let me stress that, under wave-functionalism, the wave function is treated functionally: it is defined in terms of the roles it plays in the theory. But crucially, those roles could be fulfilled by something else. As mentioned more than once, the wave function is only one possible realizer of the functions required by the law of evolution of matter. While it does have an explanatory function, it performs this role not because it is the wave function *per se*, but because it serves as a component of the dynamical law. Since these functions are multiply realizable, the wave function is not essential to fulfilling them. Therefore, although it contributes to explanation, it is not indispensable in the way a working posit must be. *What is essential is the presence of laws and a specification of the ontology of matter*. One could remain agnostic about the specific form the law takes, as implemented by the wave function, and still uphold realism about matter. Indeed, presumably one should remain agnostic about the specifics of the law, as these vary across different theories. What remains essential is the existence of laws—understood as entities that govern or describe the behavior of matter—which is a necessary working posit to explain phenomena. However, while the presence of a law persists through theory change, its details do not. Since the wave function belongs to those details, it should be regarded as a presuppositional posit, not a working one.

Discontinuity in Theory Change

The analysis provided underscores the *advantage* of my framework over wave function realism, as well as any interpretation that rejects a spatiotemporal foundational ontology. To start, let's focus on wave function realism. As discussed earlier, embracing realism about the wave function creates significant challenges for maintaining scientific realism in the quantum realm. This is because, under wave function realism, the wave function is treated as a physical field directly representing matter. From the perspective of explanationism, this means the wave function must be considered a working posit within quantum theory. The issue arises, however, from the wave function's mathematical formulation, which exists within a high-dimensional configuration space—something entirely alien to the familiar entities of classical physics. This results in an ontology that is radically different from classical mechanics, where material particles move through the familiar three-dimensional space. In wave function realism, the understanding of reality undergoes a *drastic transformation*: instead of a world full of particles situated in space and time, the world is described by a single field spread over a multi-dimensional space. Such a shift is not just a minor correction of our understanding of physical entities; it signifies a profound transformation in the spatial structure in which those entities are located. Unlike the spatiotemporal framework that I advocate, *wave function realism* does not allow any easy transition from the classical view of the world to the new quantum framework. If wave function realism holds true, there is no way to reframe the quantum theory to resemble the classical ontology, undermining any continuity between classical and quantum physics. In this context, *the working posits of quantum theory cannot be seen as a natural evolution of classical posits*. This lack of continuity obstructs any attempt to use explanationism as a defense against the pessimistic meta-induction, which raises doubts about the truth of theoretical entities that have been replaced over time.

The situation is no less problematic within *Hilbert space fundamentalism*. Like wave function realism, it treats the wave function as a working posit, but it does so without preserving any real conceptual or structural connection to classical physics. The result is the same: a radical rupture, not a gradual evolution, from classical to quantum mechanics. Similarly, *spacetime state realism* also faces issues. Though it aims to ground the ontology in physical spacetime, it still treats the wave function as a working posit, which undermines any potential for continuity and does not provide a meaningful framework for explaining how macroscopic structures constructively emerge from their microscopic counterparts. Structuralism, in this context, fails to offer a solid constructive explanation. (To be clear, as emphasized in Chapter 2, they acknowledge this and pursue structural explanations grounded in principle theories—an approach that fulfills their aims perfectly. However, for those inclined toward a constructive framework, such approaches are unlikely to be satisfying.)

Lastly, consider the *multi-field* interpretation. This approach, too, treats the wave function as a working posit. However, it might seem to offer more continuity since the multi-field is defined over spacetime, resembling the familiar three-dimensional fields of classical theories like electrodynamics. Despite this potential advantage, the introduction of the wave function as a fundamental entity undermines important symmetry properties, as discussed in Chapter 4. Although the multi-field interpretation might seem more appealing at first glance, it ultimately shares the same deficiency as the others: it *lacks the ontological and explanatory coherence* that arises from a spatiotemporal framework in which the wave function is considered a presuppositional posit rather than a fundamental entity.

In summary, while various quantum interpretations may seem to offer ways of maintaining continuity or conceptual consistency, they ultimately fail to provide the kind of explanatory and ontological unity offered by the spatiotemporal framework I propose. Treating the wave function as a presuppositional posit, as I advocate, preserves the continuity of working posits across theory change and allows for a more coherent and resilient approach to scientific realism.

Further Readings and Discussion

The references in this section provide valuable resources for understanding various philosophical approaches, especially in relation to epistemic and pragmatist views, the nature of the wave function, and other key topics in the philosophy of science. Healey (2022), along with his previous works from 2012, 2017, and 2020, offers in-depth insights into epistemic and pragmatist perspectives in the context of quantum mechanics. Critics of the pragmatist approach, such as Lewis (2020) and Jansson (2020), provide counterarguments worth considering. For a comprehensive overview of different philosophical interpretations of the wave function, Chen (2019) summarizes and evaluates various approaches.

Levin (2023) provides an overview of functionalism in philosophy, while notably Crowther et al. (2021) as well as Huggett and Wüthrich (2025) include important papers on spacetime functionalism. The ontic-epistemic distinction, a key element in understanding quantum states and their interpretation, is explored in works by Leifer (2014), Ben-Menahem (2017, 2018, 2020), Fano et al. (2019), Maudlin (2019a), López and Oldofredi (2020), Calosi and Oldofredi (2021), Hubert (2023), and Carcassi et al. (manuscript). For my own perspective on this, see Allori (manuscript b) and the references within that work.

The strategy of selective realism, also known as *Divide et Impera* or *explanationism*, has been developed by philosophers like Leplin (1997), Niiniluoto (1999), Sankey (2001), Psillos (1999), and Kitcher (1993, 2001). Additional literature exploring this strategy includes works by Stanford (2003, 2006), Chang (2003), Lyons (2006), Harker (2013), Vickers (2012, 2013, 2017), and Peters (2014). These resources offer a broad range of perspectives on key debates in the philosophy of science and provide a foundation for further research and discussion on topics such as realism, the wave function, and functionalism.

6

A Metaphysics of Bare Objects

So far, we have discussed the fundamental ontology of matter (Chapter 4) and then the wave function (Chapter 5). The fundamental ontology of matter should be microscopic, three-dimensional, and foundational in order to provide a constructive explanation that preserves symmetries. This is achieved if the wave function is understood as functionally defined, forming part of the nomological ontology—that is, as a (realizer of a given) component of the laws of nature.

In this chapter, I want to explore what appears to be a natural, though speculative, extension of this framework: *stripping away what we have so far referred to as "fundamental properties" and reclassifying them under the nomological category*, in the same way we did with spin in Chapter 4. This move leads to more complex laws, which suggests that they should be understood in terms of *structures* interconnecting objects. Nevertheless, I argue that this approach is, on the whole, more fruitful than the traditional one, as it extends more readily into the quantum domain. Moreover, it is *more parsimonious, more flexible, and altogether less problematic* than the alternatives. Note that in my account, properties such as mass and charge do not vanish entirely, but are instead regarded as effective rather than fundamental—*emergent*, like the transparency of water, from the fundamental ontology and its dynamics.

Chapter Schema Details

In the first section, I introduce and motivate my proposal, which revises what can be called the "*traditional core*"—a view endorsed by naturalized metaphysicians and realists that asserts four essential components are needed to adequately describe the world: space (or spacetime), matter, laws, and properties. I refer to this view as an *object-oriented metaphysics grounded in properties*. In contrast, I propose a perspective that requires only three ingredients: space(time), laws, and matter, without any fundamental properties. This view can be described as a *thin-object-oriented metaphysics grounded in structure*, where no intrinsic properties exist beyond those necessary for identifying the fundamental ontology. For example, if the world is composed of particles, their only property is their location in space. They possess no other fundamental properties, such as mass or charge. Their behavior is explained through structural relationships encoded in the laws of nature. More specifically, my view proposes that at the fundamental level, *there are pairs of things*: "bare" or "thin" fundamental objects (such as particles or fields)—lacking fundamental properties—and for each such object, an effective law governs its motion. These pairs form the *fundamentalia* of the world; they are what God created at the beginning of time. Thus, *there is just one kind of fundamental entity*. For instance, for a fundamental particle ontology, all particles are identical: there are no separate families like electrons, protons, or neutrons. The fundamental entities can be grouped into *effective families* (such as protons and electrons), which are distinguished by the law that governs their motion. These families are identified by the *effective law* that guides them, not their properties. So, being an electron means being a point-like particle located in space, governed by a specific effective law. This entity-law pairing is *primitive* in the sense that it is assumed without further justification. There are as many effective laws as there are families of entities traditionally considered. If particles are the fundamental ontology of matter, then they are essentially material points, guided by effective laws which together form a *complex interconnecting structure* between them describing their interactions.

After characterizing my approach, I further develop it by *comparing* it to the *traditional view and structuralism*, and argue that it ultimately provides a better framework. My proposal is the best balance between eliminating too much (structuralism) and keeping too much (traditional view). Furthermore, my framework is more fruitful, as it easily *extended to the quantum domain*. Moreover, intrinsic properties are reinterpreted as nonfundamental emerging *effective properties*—much like temperature, color, or viscosity. Thus, my approach explains why some properties are emergent while others are fundamental. It also clarifies why certain fundamental properties exist and others do not, and why the values of these properties are as they are—by eliminating the notion of fundamental properties altogether.

Finally, I address some *potential objections* to my proposal. One concern is that my view is *unnecessarily radical and mysterious*. In response, I argue that my move is not unwarranted, as traditional metaphysics relies heavily on the contentious notion of properties. Another criticism is that my approach is *conspiratorial, thus implausible*. The laws dictate the behavior of objects in such a way that they appear to possess properties, even though they do not. However, recalling that properties do exist, but they are not fundamental, resolves the issue.

Objects, Laws, and Properties

In essence, naturalized metaphysicians and scientific realists (at least prior to the advent of quantum mechanics) essentially believe that fundamental entities exist in space and time and evolve according to the laws of nature, as determined by their intrinsic properties. This traditional way of understanding explanation in physics can be summarized as consisting of four elements: *spacetime, fundamental objects, their intrinsic properties, and the laws of nature*. The fundamental entities could be, for example, particles in classical mechanics, or both particles and fields in classical electrodynamics. The difference in their fundamental nature is reflected in their mathematical representations: particles are represented as three-dimensional points, while fields are described as functions assigning a field value to each point in space. It is often said that laws of nature govern or describe the behavior of these entities: Newton's law dictates the trajectories of particles, and Maxwell's equations describe the evolution of electromagnetic fields. However, more than just laws are needed to fully capture the behavior of fundamental entities; their intrinsic properties must also be specified. For instance, in classical mechanics, particles must have mass in order to explain why different bodies behave differently under the same conditions, as described by Newton's law of gravitation. If body 1 and body 2 have respective masses m_1 and m_2, Newton's law dictates that body 1 accelerates toward body 2 with acceleration $a_{12} = \frac{F_{grav}}{m_1}$, while body 2 accelerates toward body 1 with $a_{21} = \frac{F_{grav}}{m_2}$. Similarly, in classical electrodynamics, charge is necessary to explain experimental outcomes. If particle 1 and particle 2 have the same mass, but particle 1 turns left in a magnetic field while particle 2 turns right, this divergence can be attributed to their opposite charges. Thus, traditionally, particles are classified into *distinct families based on their intrinsic properties*. For example, a particle is identified as an electron or a proton due to its particular mass and charge. These properties are considered definitional of the object, meaning they are intrinsic to it. An electron, for instance, is defined by its mass $\left(m_e = 9.11 \times 10^{-31} kg\right)$ and charge $\left(q_e = -1.60 \times 10^{-19} C\right)$. In this sense, fundamental objects are "thick"—they "wear" their intrinsic properties, which are also seen as fundamental. These properties must be specified to provide a complete description of the world. On this basis, fundamental objects are grouped into families that share the same intrinsic properties: all electrons have the same mass and charge; they are identical in this respect, even if they can be labeled differently based on their location.

From this perspective, all fundamental entities follow the *same laws of motion*. Fields obey Maxwell's equations, and particles follow Newton's second law $(F = ma)$, supplemented by the gravitational and electromagnetic force laws (F_{grav} and F_{EM}), such that the total force is $F = F_{grav} + F_{EM}$. The same law applies differently to entities in different families because of their varying properties. According to this ontology of matter—which consists of fundamental objects—there is *one set of fundamental laws and many fundamental intrinsic properties*, leading to the existence of *many different families* of fundamental objects. It is as if the guiding principle is to minimize the number of fundamental laws (e.g., Newton's second law, the law of gravitation, Maxwell's laws of electrodynamics) and to add fundamental properties as needed to ensure empirical adequacy. This, in turn, necessitates multiplying the number of fundamental kinds of entities (families) by introducing additional fundamental intrinsic properties.

In conclusion, this traditional understanding of the interplay between objects, properties, and laws can be described as an *object-oriented metaphysics grounded in properties*, requiring four categories of fundamental things: spacetime, objects, properties, and laws.

The Ingredients of the Traditional View

Many debates are associated with the traditional metaphysical understanding, each linked to one of its fundamental ingredients: space and time, laws of nature, material objects, and properties.

First, questions arise regarding the *nature of space and time*: is space a substance, or does it merely consist of relations between objects? These foundational inquiries set the stage for deeper metaphysical analysis.

Equally central is the debate over *the nature of laws*. According to Humeans, laws are simply the axioms and theorems of the best descriptive theory of the world. In contrast, broadly understood necessitarians argue that laws exist over and above physical events, with primitivists or anti-reductionists maintaining that laws belong to their own metaphysical category. This issue becomes especially pressing when we consider how laws are supposed to govern material behavior. For Humeans, laws do not govern at all: they are merely part of the "mosaic" of facts they are supposed to summarize, which compromises their explanatory power (Armstrong 1983). On the other hand, non-Humean accounts propose that laws do govern, but this raises a further metaphysical question: How can laws influence the behavior of material entities if they are not themselves located in space and time (Loewer 2012)? The problem with laws is also apparent in the deductive-nomological model of explanation (Hempel 1952), according to which explanations are achieved by subsuming phenomena under laws. This model has been heavily criticized for failing to account for the asymmetry of explanation—our intuition that effects are explained by causes, but not the other way around. Because of such limitations, some philosophers have proposed that all explanatory work be reassigned to properties, particularly *dispositional properties*, thereby eliminating the need for laws altogether. In this view, objects possess dispositional properties that are interpreted as causal powers which bring about events (Cartwright 1983).

In this context, the notion of *material objects* also becomes controversial. Descartes claimed that the fundamental property of material objects is their being located somewhere. But what does this require, metaphysically speaking? Newton defined mass as a measure of the quantity of matter in an object, but using mass as a definitional criterion for material objects would exclude massless particles like photons. To address this, Wilson (2006) characterizes material objects in a minimal yet effective way: fundamentally, they are entities that are not mind-dependent.

Properties themselves are a central point of contention. They are typically understood as universals, in contrast to objects, which are particulars. According to the traditional core, properties such as "having mass equal to $9.11 \times 10^{-31} kg$" are instantiated by all electrons: they both exemplify and share such a property. Adherents of the traditional view are realists about properties—they argue that properties exist, are mind-independent, and play an essential role in the operation of laws and in explaining phenomena. Nominalists, on the other hand, deny the existence of such abstract entities. As a result, the debate over intrinsic properties has gained significant traction.

A major point of contention is the *distinction between categorical and dispositional properties*. Categorical properties are those that an object possesses in virtue of being a certain way, while dispositional properties concern an object's abilities or tendencies to act, or to be acted upon, in certain ways. There are varying positions in this debate: some philosophers argue that all properties are purely categorical, others that all are purely dispositional, while still others propose hybrid views. The hybrid approaches either posit that both types of properties exist, or that each property possesses both categorical and dispositional aspects. However, dispositional properties themselves have been questioned. If categorical properties describe the present state of a property bearer, dispositional properties concern what the bearer will do or become if certain conditions are met. Thus, it is argued, dispositional properties do not actually describe the current status of the object, but rather its potential future behavior (Goodman 1955). This calls into question their status as genuine properties. Furthermore, the very definition of what counts as dispositional or categorical is controversial (Cross 2005). One may claim that a property is dispositional when its essence consists in what it prescribes its bearer to do, whereas a categorical property is defined differently: either by its essence being independent of what its bearer does, or by its persistence across all possible worlds. It remains unclear whether these definitions are mutually compatible. Examples help clarify the debate. Solubility is typically seen as a dispositional property: a substance is said to be soluble if, when placed in water, it dissolves. On a more fundamental level, mass and charge can also be understood dispositionally, since they determine what a particle is disposed to do under the influence of forces. Yet mass and charge are also viewed as categorical by some, because they are taken to be definitive of what a particle—say, an electron—is in all possible worlds. Spatial location and relations are generally cited as examples of categorical properties. Furthermore, within the category of dispositional properties, a distinction is made between causal powers—such as mass, which are thought to be actively involved in causal processes—and propensities, such as the half-lives of radioactive elements, which describe tendencies whose manifestations do not depend on specific circumstances. These are referred to as "absolute" dispositions.

In light of these interconnected debates, I believe that the traditional metaphysical framework—grounded in spacetime, material objects, laws of nature, and intrinsic properties—rests on *deeply contested foundations*. Each of its core elements invites serious philosophical scrutiny, from the ontological status of laws and the nature of space and time to the very identity of material objects and the classification of their properties. These tensions suggest that any robust metaphysical account of the natural world must either significantly revise or move beyond the traditional paradigm, addressing its foundational assumptions.

Quantum Traditional Metaphysics?

Beyond the challenge of adequately defining properties, the most serious obstacle faced by the traditional metaphysical approach lies in *its application to the quantum domain*. At first glance, it may appear that quantum theory, too, relies on properties to account for different behaviors under identical conditions. Indeed, quantum theory seems to introduce *more properties*, such as spin, to explain the behavior of various particle types. For example, bosons are defined by their integer spin values, while fermions possess half-integer spin values.

However, the situation must be analyzed differently depending on the interpretive approach one endorses. To begin with, within wave function realism, one cannot appeal to fundamental particle properties to explain behavior in the way the traditional model allows. For instance, in the traditional view, the differing behavior of electrons and protons in a magnetic field is attributed to their opposite electric charges. But wave function realism denies the fundamentality of both particles and their properties. Instead, matter is described wholly by the wave function, whose only fundamental feature, if any, is arguably its field value. In Albert's formulation of wave function realism, three-dimensional material entities are merely functional shadows of the high-dimensional wave function. Thus, wave function does not naturally delineate specific properties to assign to each of these shadows in any obvious or consistent way. All matter is described by a single, universal wave function with a uniform field value. When we observe electrons rising in a magnetic field, Albert's account does not explain this in terms of the particle's charge. Rather, it posits that some aspect of the wave function plays the functional role typically associated with an electron, and under those conditions, that role manifests as upward motion in the magnetic field. As we saw, Ney offers a similar interpretation. According to her, electrons exist only in a derivative sense: they are approximations that emerge in situations where the behavior of the wave function can be effectively captured by positing electrons. These emergent entities appear to have charge only because they behave in a way that aligns with how charged particles are expected to behave—for instance, rising rather than falling in a magnetic field. That behavior is the result of what a certain segment of the decomposed wave function does under specific conditions in three-dimensional space. In both cases, then, particles and their properties are not fundamental constituents of reality. Rather, they emerge from the wave function—whether functionally, structurally, or through symmetry and partial instantiation. The notions of objects and properties, while practically useful, do not reflect any underlying metaphysical structure. They are conceptual tools, arising from and dependent on the behavior of the wave function. Thus, *in wave function realism, properties assigned to particles play no fundamental explanatory role in quantum phenomena*. Since particles and their properties are not fundamental, they cannot serve as explanatory anchors. Talk of particles and their features serves as a convenient shorthand, but it does not reveal the true nature of reality. Instead, both particles and their properties are among the things to be explained, not the basis of explanation. This stands in stark contrast to the traditional approach, which appears increasingly ill-suited to account for the nature of quantum reality under wave function realism.

Similar conclusions arise in the contexts of spacetime state realism and Hilbert space fundamentalism. In spacetime state realism, properties traditionally ascribed to particles—such as charge or spin—are not fundamental, but rather emerge from the overarching state of the spacetime region. This perspective aligns with the view in wave function realism, where particle properties are not fundamental explanatory entities but are instead emergent from the structure and evolution of the wave function. Within Hilbert space fundamentalism, the entire structure of reality—including space, particles, and fields—emerges from the properties and dynamics of a fundamental vector in Hilbert space. Consequently, entities such as particles and their associated properties are not fundamental, but are emergent phenomena arising from the underlying Hilbert space structure.

In contrast, the *multi-field* approach offers a different perspective. Developed within the context of pilot-wave theory, it seeks to interpret the wave function as a *poly-wave* in three-dimensional space representing a physical, rather than nomological, entity. Therefore, the wave function in this framework cannot "absorb" properties, which are thus traditionally understood as fundamental features of the physical world. As such, the multi-field approach does not support the notion that particle properties are nonfundamental. It *aligns more closely with traditional metaphysics*, even if it faces the difficulties discussed in Chapters 3 and 4.

Toward a New Understanding

My approach seems to be compatible with the traditional schema: matter moves through space and time according to a law, guided by the fundamental properties of objects. However, as I explain next, extending this schema to all properties—including spin—leads to some unwelcome consequences, as anticipated in Chapter 4. In other words, under the framework I propose, certain properties—most notably spin, the quintessential quantum property—are better understood as aspects of the wave function rather than as intrinsic features of particles (we have seen this in Chapter 4). This is due to the *contextual nature of spin*: its value depends on how the measurement is conducted. A true or genuine property should reflect a central, perhaps essential, characteristic of an object—something definitional that remains constant regardless of external conditions. If we assume that the act of measurement accurately reveals the property in question, then its value should not vary depending on the measurement context. In contrast, if a property's observed value shifts with the circumstances of measurement, then it is either not a genuine property, or the measurement process fails to reveal the property itself—merely documenting how the system is altered by its interaction with the apparatus.

This naturally raises a broader question: *Should all other properties be regarded in a similar light*? One motivation for doing so lies in the *problematic nature of properties* themselves, as just outlined. Another stems from a concern for *ontological economy*: if all explanatory work can be done by appealing to laws alone, what role remains for properties? Eliminating properties would, at the very least, reduce the number of philosophical puzzles we face. This sort of eliminativism is not without precedent. It is a hallmark of structuralist approaches, which are typically motivated by commitments to simplicity and parsimony (as examined in Chapter 2). However, their empiricist stance—responding to the threat of underdetermination by eliminating not only properties but also objects—goes too far. As I argued earlier in Chapter 2, preserving objects is crucial for sustaining the possibility of constructive explanation.

Therefore, *instead of eliminating objects, my proposal involves removing properties from the fundamental level.* This avoids the radical consequences of structuralism. On my view, the explanatory power in constructive explanations derives entirely from laws—not from any intrinsic properties of objects. The behavior of fundamental entities is fully governed by their dynamics, which reflect the laws they follow. Moreover, macroscopic objects are conceived as composites of microscopic constituents, like elements in a Lego set. *Macroscopic properties*—such as the transparency of water—are understood as the result of interactions among these constituents, interactions governed by laws. *Property-talk may be used as a convenient shorthand*, but it does not deepen our metaphysical understanding. If macroscopic properties can be conceptually absorbed into the laws themselves, then their independent existence appears unnecessary. In this light, why not extend this reasoning to *intrinsic properties as well*? These, too, could be seen as effective, emergent features, rather than as fundamental. That is, while properties such as mass, charge, or spin may still be invoked in scientific discourse or everyday explanation, they no longer play a metaphysically grounding role. Instead, they are understood as *higher-level constructs that emerge from the interaction of more basic elements*—namely, the fundamental objects and the laws governing their behavior. For example, what we call "mass" can be seen as a way of capturing certain regularities in motion and interaction—regularities that ultimately stem from the application of specific dynamical laws, rather than from any intrinsic essence possessed by the particle. Similarly, "charge" or "spin" may function as practical descriptors of predictable patterns in experimental outcomes, without requiring that there be any underlying metaphysical property corresponding to them. These effective properties exist as tools for navigating the macroscopic world and organizing our understanding of phenomena, but they do not point to fundamental constituents of reality. *Their explanatory utility does not imply ontological fundamentality.* Therefore, these properties can be seen as derivative, arising from the structure and function of the laws themselves, rather than as fundamental features of objects. Recognizing this allows us to retain the practical use of property-talk without inflating our ontology—preserving ontological parsimony while maintaining explanatory power.

Adopting a structuralist intuition, *my framework inverts the dispositionalist view*: rather than reducing laws to the powers or dispositions of properties, I propose *absorbing properties into the laws.* The next section explores this approach in more detail. My ontology posits just *two fundamental components: the fundamental material entities and the effective laws associated with them.* In this view, all features traditionally regarded as fundamental properties—mass, charge, spin, or even the category of matter itself—are not properties at all. Crucially, while this framework dispenses with properties at the fundamental level, it does not abandon objects. In the final section, I compare my stance with structuralism: while both views eliminate properties, my view preserves the existence of fundamental objects.

Thin Objects and Effective Laws

My position calls for a foundational rethinking of the status and role of the laws of nature, which I will later suggest are best interpreted within a structuralist framework. At the core of this view is the idea that, much like in the traditional picture, the world consists of spacetime and fundamental entities moving around in it. However, unlike the traditional account, these entities possess *only one fundamental property*—the single feature that uniquely defines their nature (see also Allori 2023d). For instance, if the fundamental ontology consists of particles, then the only property that characterizes them is their spatial position. To be a particle, in this sense, is simply to be located at a specific point in space at a given time. Alternatively, if the fundamental ontology is that of fields, then the defining property is the assignment of intensity values at each spatial point over time. Being a field, then, amounts to having a value defined across a spatial domain. This stands in sharp contrast to the traditional understanding—outlined in the previous section—which posits a diversity of fundamental entities distinguished by multiple intrinsic properties and sorted into different kinds or families. On my view, by contrast, *all fundamental entities are ontologically identical*: there is only one category of entity. So, in a particle-based ontology, we do not distinguish between electrons, protons, or neutrons based on intrinsic features. All such distinctions collapse into the singular notion of particles as simply being "full" points in space (as opposed to "empty" ones). Crucially, the idea of *fullness* is primitive—it cannot be further analyzed except by saying that such points are *full of matter*.

A likely objection to this proposal is that efforts to describe matter without appealing to intrinsic properties have been attempted before and have failed. The standard argument is that we need properties to explain why entities behave differently under identical conditions. For example, if electrons and protons lacked distinct intrinsic features, how could we explain why only electrons deflect upward in a magnetic field? Indeed, many have taken intrinsic properties to be so essential that they have tried to reduce laws to properties. In response, I argue that it is possible to account for these behaviors without assuming intrinsic properties. On my view, differing behaviors arise not because entities possess different properties (they do not), but because they are subject to different *effective laws*. These effective laws specify the various ways in which a general law applies to the fundamental entities. Thus, what appear in the traditional framework as *different kinds* of entities are reinterpreted as the *same kind* of object, governed by *distinct effective laws*. To illustrate: in a particle ontology, what we customarily describe as different particle species—electrons, protons, etc.—are in fact a single type of particle governed by multiple effective laws. As already noted, this approach sharply contrasts with dispositionalism, which shifts all explanatory responsibility onto properties and seeks to eliminate laws. My framework takes the opposite route: it transfers explanatory force from properties to laws. As such, this view can be seen as a variant of "nomic nominalism," in which laws constitute the entirety of the theoretical architecture, while properties are eliminated at the fundamental level. Under this picture, a statement like "has mass $9.11\times10^{-31}kg$" is redefined: it simply marks membership in the class of entities governed by the effective laws associated with that parameter. Rather than positing a plurality of fundamental properties, I propose that we posit a single kind of fundamental entity, accompanied by as many effective laws as are needed to align with empirical phenomena. The structure linking these entities is encoded in the general laws, which are then instantiated through the differentiated action of the effective laws. I will now proceed to develop this idea further.

Constants and Variables

Let us turn to classical physics, particularly Newtonian mechanics and classical electrodynamics. Within these frameworks, the fundamental forces—such as gravity and electromagnetism—are governed by laws that include fixed, universal constants. Consider Newton's law of universal gravitation: $F_{rav} = G\frac{m_1 m_2}{r^2}$. Here, G, the gravitational constant, has a fixed value: $G = 6.6743\times10^{-11}m^3kg^{-1}s^{-2}$. Constants like G are integral to the formulation of the law itself—they remain the same across all applications, independent of the specific objects involved. In contrast, parameters such as mass and charge differ from one object to another. They are not built into the law's universal structure but instead characterize the particular systems to which the law is applied. That is, *parameters describe matter*, not the law. This distinction is generally uncontroversial: mass and charge are conventionally treated as intrinsic properties of physical entities. As previously discussed, the traditional metaphysical picture explains empirical phenomena by ascribing such properties to particles in addition to their spatial positions. This results in a taxonomy of fundamental particle kinds—electrons, protons, neutrons, and so forth—each defined by a unique bundle of intrinsic features.

Now imagine we reframe this perspective. Suppose we no longer regard parameters like mass and charge as intrinsic properties of matter, but rather *treat them as part of the law*'s specification, alongside constants like G. In this reinterpretation, these parameters are absorbed into the laws themselves, becoming part of the dynamical framework rather than features of the objects it governs. Properties, in this view, are effectively "in the law"—what previously seemed intrinsic to particles is now seen as a feature of the governing dynamics. Adopting this perspective implies that empirical adequacy is no longer achieved by assigning intrinsic properties to entities. Instead, it is achieved through the formulation of appropriate *effective laws*. The explanatory work shifts from the ontology of objects to the structure and function of laws, thereby challenging the necessity of intrinsic properties in our understanding of physical reality.

Let us illustrate this with a *concrete example*. In the traditional view, three fundamental particle types are distinguished: electrons, protons, and neutrons. Corresponding to these are six parameters—three masses m_e, m_p, m_n and three charges $q_e, q_p,$ (with obvious notation). Let us set aside charge for the moment and focus solely on gravitational interaction. Newton's law, when applied to an *electron*, can be rewritten as $F_{'electron'} = \frac{H_1}{r^2}$, where $H_1 := Gm_eM$, with M representing the mass of the second body producing the gravitational field. The form of the law—its structure—remains invariant across particle types: it maintains a $1/r^2$ dependence. However, each particle type has its own constant H_i, determined by its specific mass. What were once interpreted as three distinct kinds of particles can now be reimagined as a single kind of particle governed by three distinct effective gravitational laws: $Eff\ law_1 = \frac{H_1}{r^2}$ with $H_1 = Gm_eM$; E $Eff\ law_2 = \frac{H_2}{r^2}$ with $H_2 = Gm_pM$; $Eff\ law_3 = \frac{H_3}{r^2}$ with $H_3 = Gm_nM$. Yet the situation is more nuanced. The value of H_1 for any given particle depends on the presence of other particles. If there is only one particle (say, particle 1 with mass m_1), its effective law becomes $H_1 = Gm_1M$. But since there is no second body, $M = 0$, and thus $Eff\ law_1 = 0$. If a second particle (particle 2 with mass m_2 (written as m_1 above)) is present, the effective law for particle 1 becomes $Eff\ law_1 = \frac{H_1}{r^2}$, where $H_1 = Gm_1m_2$. With a third particle added, the effective law becomes a sum of interactions: $Eff\ law_1 = \frac{H_{1,3}}{r^2} + \frac{H_{1,3}}{r^2}$, where $H_{1,2} = Gm_1m_2$, and $H_{1,3} = Gm_1m_3$, and so on for systems with more particles. *This framework generates a complex, interdependent network of effective laws that dynamically link all particles.* The resulting structure aligns closely with the structuralist notion of interrelated dynamical roles, rather than intrinsic characterizations.

To recap: in my approach, the fundamental material ontology of a theory consists of the spatiotemporal microscopic entities which allow for a constructive and symmetry preserving explanation of the phenomena. In classical mechanics, these entities are particles. Crucially, in my view there is only one kind of fundamental particle, and its only defining feature is spatial location, since this is what defines its corpuscular nature (as opposed to, say, field-based). Differences in behavior under identical conditions, which in the traditional account are attributed to differences in intrinsic properties, are here explained by the fact that these otherwise identical entities are governed by *different effective laws*. This becomes perhaps more intuitive when we turn to classical electrodynamics. Suppose that, in the traditional view, electrons are represented as dark dots, protons as gray dots, and neutrons as white dots, each tracing its own trajectory through spacetime according to the same law. In my view, by contrast, there are no such colors. There are only indistinguishable points, each following a different path because each is governed by its own effective law (see Figure 6.1):

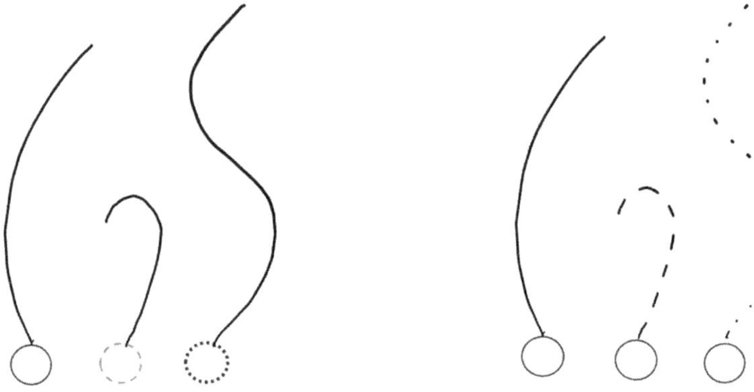

Objects, laws and properties

Thin objects, effective laws

Figure 6.1 Thick vs. thin objects.

- The electron is a point governed by $Eff\ law_1$, and thus traces what was traditionally the "dark" path.
- The proton follows $Eff\ law_2$, moving along the "gray" (dashed) path.
- The neutron follows $Eff\ law_3$, taking the "white" (pointed) path.

All particles are ontologically identical; only their dynamics differ. At the fundamental level, the full ontology O consists of pairs $\wp$ of material entities M and their governing effective laws $Eff\ law$: $\wp = \{M, Eff\ law\}$. In the traditional framework, particles carry properties and are subject to a single overarching law $\mathcal{L}$: $O = \{Matter; properties; law\} = \{electron\, m_e, q_e, \mathcal{L}; proton, m_p, q_p, \mathcal{L}; neutron, m_n, q_n, \mathcal{L}\}$. By contrast, in my view, we have three ontologically indistinct particles, each paired with its own effective law: $O = \{\wp_1; \wp_2; \wp_3\} = \{(M, Eff\ law_1); (M, Eff\ law_2); (M, Eff\ law_3)\}$ All properties are embedded in the laws themselves, and particles are not defined by intrinsic traits but by the dynamical laws that govern them. Each particle's association with its effective law—its pairing $\wp$ —is taken to be *primitive*.

It may be helpful to *revisit the orchestra analogy* introduced in Chapter 5, particularly in relation to how it illustrates the role of the wave function in the framework under discussion (Figure 6.2). In that chapter, the analogy was used to clarify the nomological role of the wave function. According to the traditional metaphysical view, matter is analogous to the musicians in an orchestra, while different families of particles correspond to different groups of musicians playing distinct instruments. The laws of nature act like the musical score, guiding how the musicians (particles) interact. In this analogy, the score that the conductor sees—representing the full symphony—is the same for the entire orchestra, but each "family" of musicians follows its own

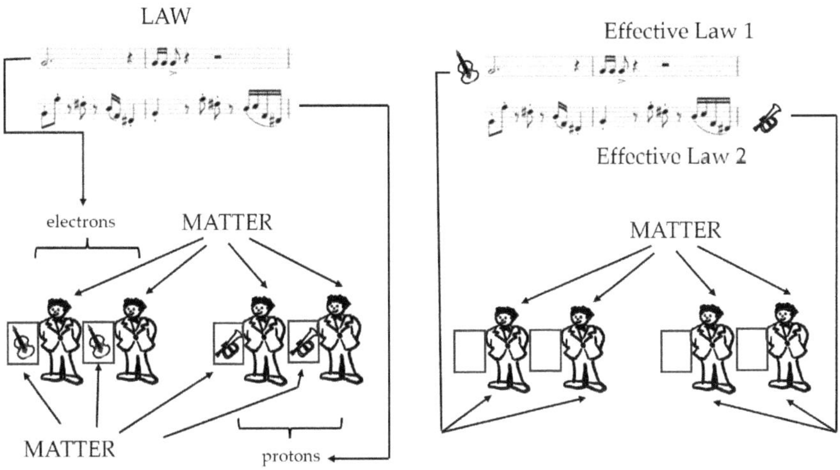

Figure 6.2 The orchestra analogy: the traditional view (left) and my account (right).

specific part. For instance, violins play one section of the score because they are violins, not clarinets. In my framework, however, the picture shifts. Matter is still likened to musicians, but with a significant change: all musicians are identical, and all play music in the symphony. There is no longer a distinction between different families of musicians, each playing a specific instrument. Instead, there is a single set of musicians, all of whom participate in the performance. The law of nature—still represented by the full score—is divided into parts, but these parts are assigned to individual musicians based on their roles, rather than on any inherent distinction like being a violin or clarinet. Some musicians read what we might think of as the "violin part," others the "clarinet part," but this division does not arise from a difference in kind among the musicians. Rather, it arises from a primitive pairing: each musician is simply assigned a particular section of the score to follow. *Why is this particular pairing chosen over another*? The answer, within this framework, is that it just is—this is how the structure of the theory is set up. The key difference between this view and the traditional one lies in how the relationship between musicians and the score is understood. In the traditional account, each instrument is assigned a specific part because of what it intrinsically is—a violin, a clarinet, etc. In my view, by contrast, there are no such intrinsic distinctions. The pairing between a musician and a part of the score is not grounded in any fundamental difference in nature. It is simply taken as primitive. One might then ask: Why is this approach more problematic than the traditional one, in which each instrument is assigned its part based on its identity? The answer lies in how we conceive of particle identity and role. Rather than assigning a distinct role based on what a particle *is*, roles are assigned without assuming any intrinsic difference between the particles. Identity arises not from categorically distinct traits, but from the dynamical role a particle plays within the system.

It is important to remember that *this view does not amount to eliminativism regarding properties.* I am not denying the existence of properties altogether. Rather, I am proposing a shift in their ontological status: properties do exist, but they are not fundamental features of reality. They emerge from a deeper level of structure—specifically, from the dynamics governing the fundamental ontology. As suggested earlier, the familiar way we speak about particles and their properties—"property-talk"—is preserved, but only as a practical and effective way of describing what emerges from a more basic level. Macroscopic properties such as temperature, pressure, transparency, or fragility arise from the behavior and interactions of microscopic constituents. They are effectively functions of the underlying fundamental matter and the laws that govern its evolution. For example, temperature is not a fundamental quantity, but a statistical manifestation of the kinetic activity of particles. Similarly, transparency or fragility reflects regularities in how microscopic structures respond to external conditions. The idea is that just as macroscopic properties are not fundamental, the properties traditionally considered fundamental to particles may not be either. Even properties traditionally regarded as intrinsic—those thought to define the very nature or essence of matter—are reconceptualized in this framework. Instead of viewing them as fixed, categorical attributes of objects (such as mass, charge, or spin), I argue that these properties are best understood as aspects of the dynamical roles that objects play. They are encoded in, and must be interpreted through, the behavior that emerges from the laws governing the system. In this sense, what appears to be an intrinsic feature is actually a reflection of how the object behaves under certain conditions, as dictated by its effective laws. Thus, while properties are not fundamental, they are not dismissed. They are reinterpreted: they arise from, and are defined by, the interplay of matter and dynamics at the most basic level. This view *retains the explanatory power and utility of properties*, but integrates them into a structurally grounded, property-less metaphysics.

Symmetries, Laws, and Structure

Let me open a parenthesis about something we have already touched on, namely the role of symmetry. As previously discussed, preserving symmetries in physical theories provides a compelling reason to include only the most basic entities within the fundamental ontology of matter. In particular, electromagnetic fields do not qualify as fundamental in this context and therefore should not be considered part of the basic building blocks of matter. However, what I would like to emphasize now is that *symmetry considerations can also yield deeper metaphysical insights.*

As we saw already in Chapter 4, electromagnetic fields are not foundational, so they are better regarded as part of the law, rather than as part of the fundamental ontology. This is because otherwise the theory loses symmetry properties. In fact, for instance, consider the case of time-reversal symmetry. One way to maintain this symmetry in classical electrodynamics—and to account for how magnetic fields transform under such symmetry—is by attributing a different mathematical nature to them. Specifically, magnetic fields can be understood not as standard vectors, but as axial vectors or pseudo-vectors (as discussed in Arntzenius 2004 and Malament 2004). By definition, a pseudo-vector behaves like a regular vector under spatial rotations but reverses its sign under operations like spatial reflections or time reversal. Hence, the magnetic field's peculiar transformation behavior follows from its pseudo-vector nature—it is intrinsic to its mathematical role. But this leads to a further question: if the magnetic field looks like a vector, why should we believe it is, in fact, a pseudo-vector? To adopt such a view, we must provide independent justification for this ontological reassignment.

This is where my approach offers a natural resolution. If we treat electromagnetic fields not as physical entities but as part of the laws of nature—much like the wave function in certain interpretations of quantum mechanics—then there is no need to assign them a physical nature at all. If so, there's no need to explain what they are or why they transform in a particular way. Instead, they simply behave as the symmetry of the theory requires them to because of their nomological, rather than ontological, role. Accordingly, if symmetry preservation demands certain transformation rules, those rules define how the fields behave—not as objects, but as features of the law. In this framework, we can define electric and magnetic fields, respectively, as a true vector and a pseudo-vector, consistent with symmetry requirements (as argued in Allori 2015a).

Let's explore how the reconceptualization in this chapter helps eliminate the need for electromagnetic fields by folding them into the law. Reconsider the previous example, this time focusing on the Coulomb interaction, which is a simplified case of the electromagnetic force. The Coulomb law takes the form $F_{Coulomb} = k\frac{q_1 Q}{r^2}$, where $k = 8.99 \times 10^9 \, kg\, m^3\, C^{-2}\, s^{-2}$ is Coulomb's constant. This law describes the force exerted by a particle with charge Q on another particle with charge q_1. From here, the electric field is traditionally defined as $E = k\frac{Q}{r^2}$, which allows us to express the Coulomb force as $F_{Coulomb} = q_1 E$. Now, consider the case where particle 1 is an "electron." We can rewrite the force as $F_{electron} = q_e E = \frac{K_1}{r^2}$, with $K_1 \stackrel{\text{def}}{=} kq_e Q$. Just like in the gravitational case discussed earlier, the structure of the law—its mathematical form—remains consistent across all particles; it is always an inverse-square law. However, the proportionality constant K_i varies depending on the particle in question, effectively giving rise to different versions of the law for different particle types. That is, we obtain a set of effective laws: $Eff\ law_1 = \frac{K_i}{r^2}$, with $i = 1,2,3$ indexing what were traditionally considered different fundamental particle families. So, in this reformulation, *not only do we dispense with the notion of particle families, but we also eliminate the electromagnetic fields themselves from the fundamental ontology.* The fields are no longer treated as physical entities but are instead absorbed into the definition of the effective law, as we urged to do already in Chapter 4. We get one effective law for the "electron" ($Eff\ law_1 = \frac{K_1}{r^2}$, where $K_1 = kq_e Q$); one for the "proton" ($Eff\ law_2 = \frac{K_2}{r^2}$, where $K_2 = kq_p Q$); and one for the "neutron" ($Eff\ law_3 = \frac{K_3}{r^2}$, where $K_3 = kq_n Q = 0$, because the neutron is electrically neutral).

From this perspective, *symmetry principles become a guiding tool for ontological decisions*. As discussed in Chapter 4, they help us identify which elements in a theory are foundational—i.e., which represent matter itself, rather than part of the nomological structure. If including a certain object in the fundamental ontology leads to a symmetry violation, it suggests that the object in question is not foundational. Therefore, it is better to think of it as not belonging to the ontology of matter, but rather to the laws governing it. In this way, symmetry transformations offer more than just mathematical consistency; they serve as metaphysical indicators, distinguishing between ontological and nomological entities.

Moreover, *symmetry considerations might even help identify the laws themselves*. Combined with principles of simplicity, one might argue that symmetries serve as generative constraints for the laws that govern these fundamental objects. While there is no rigid algorithm for constructing a physical theory, history shows that many fundamental laws have been introduced precisely through appeals to symmetry. Consider classical electrodynamics again: while fields are introduced to achieve empirical adequacy, their transformation behavior must still preserve symmetries, like Lorentz invariance. A similar logic applies in quantum mechanics. For example, in the pilot-wave theory, Dürr, Goldstein, and Zanghì (1992) formulated the guiding equation by asking for the simplest law that is invariant under rotations, Galilean boosts, and time reversal, and which treats physically equivalent wave functions consistently. Symmetry principles were likewise instrumental in the development of relativity theory, particularly in conjunction with the constancy of the speed of light. Figures like Wigner and Weyl were among the earliest to recognize the foundational role of symmetry groups in quantum mechanics, a practice that continues today in areas such as string theory, where dualities are a key tool. Of course, one should be cautious: the historical and epistemological role of symmetries in theory construction does not automatically settle questions of ontology. Still, if symmetry constraints could, in principle, determine both the form of the law and the type of objects involved, that would be a major step toward a unified explanatory structure. Symmetry would then not just help articulate the law, but also select the ontology and, in turn, explain the appearance of physical properties as encoded in effective laws.

On another note, let me explore *the nature of the laws* in this framework. In short, in my view the fundamental ontology consists of thin, identical entities that are individuated solely by their intrinsic nature; the law of nature then governs how these entities evolve. As we saw, to ensure empirical adequacy, the law must fragment into a set of effective laws, each associated with a distinct subset of these fundamental objects. These effective laws include parameters that correspond, in the traditional framework, to the properties we ascribe to particles. Thus, we end up with a *relational nomic structure*: a network of effective laws connecting the fundamental entities. This naturally raises several philosophical questions, particularly regarding ontological priority: Does the fundamental law exist independently of the objects it governs? Do the effective laws stand on their own, or are they reducible to the fundamental law? These questions belong to the broader metaphysical discussion on the nature of laws—whether one adopts a Humean stance, where laws supervene on the distribution of matter, or a non-Humean view, where laws are primitive. Fortunately, we can sidestep that debate entirely by adopting a structuralist position. From this perspective, regardless of whether one sees laws as metaphysically dependent on matter or not, the combination of thin objects and their associated effective laws constitutes a relational structure. This structure, whether fundamental or emergent, both describes what matter *is* and determines how it *behaves*.

Extension to the Quantum Domain

In this section, I aim to show how the perspective I've developed *extends naturally into the quantum realm*. As discussed, my view holds that matter is described by microscopic spatiotemporal variables without intrinsic properties. The relatively smooth transition from classical to quantum mechanics supports the idea that this framework applies in the quantum domain as well—with one key difference: matter now evolves according to a law that incorporates the wave function.

Consider, for example, the pilot-wave theory. In this framework, matter consists of particles, and their dynamics are governed by a law that includes constants like the reduced Planck's constant and parameters such as mass. According to the view I propose in this chapter, where objects are thin and lack intrinsic properties, these parameters are considered part of the law itself. Thus, what we conventionally refer to as effective laws arise when these parameters are treated as fixed constants. In this way, the fundamental law can be understood as fragmenting into a family of effective laws, each tailored to a specific empirical context by the inclusion of different parameter values. Since fundamental entities must be both foundational and spatiotemporal, the wave function is excluded from the basic ontology of matter. Therefore, the wave function is best interpreted in functional terms, as suggested by the wave-functionalism approach of Chapter 5.

Interestingly, as anticipated earlier in this chapter, the account I defend is also *compatible* with a wave function realist framework, as well as spacetime state realism and Hilbert space fundamentalism. In my proposal, traditional properties are eliminated, and the explanatory role once performed by them is now carried out by the laws themselves. These laws, rather than properties, become the primary explanatory tools—they govern the evolution and behavior of the fundamental constituents of reality. Focusing on wave function realism, as previously examined in the cases of both Albert and Ney, references to properties are essentially fictional. The familiar three-dimensional objects of our everyday experience, within their respective frameworks, are either functionally characterized or privileged through symmetry. In neither case, however, is there any appeal to intrinsic properties in the traditional metaphysical sense. In Albert's account, to be a system of three-dimensional objects is simply to be governed by a certain Hamiltonian; in other words, such a system's identity is bound up with its behavior—specifically its position and inter-particle relations in three-dimensional space. Ney, by contrast, appeals to symmetry arguments to single out three-dimensional configurations as special, with these configurations emerging derivatively from a high-dimensional wave function. Again, no fundamental or derived properties are postulated in any substantive sense. Macroscopic and intrinsic properties suitably emerge, just as in spacetime state realism and Hilbert space fundamentalism.

To be clear, these positions are distinct from the structuralist metaphysics grounded in thin objects that characterizes my framework. In my approach, there is a *stronger continuity*: the thin objects are situated in three-dimensional space, and the structure that governs them—the dynamical relations among these entities—is provided by the wave function, which itself is not regarded as a material field or substance. By contrast, the ontology in wave function realism and similar views functions somewhat differently. In these approaches, the wave function is considered the very material basis of the world. It lacks fundamental properties except for the amplitude it assigns to points in configuration space, in the case of wave function realism. As such, the wave function itself can be seen as a kind of thin object. However, this entity does not directly explain empirical phenomena. Observable features of the world are instead accounted for by derivative three-dimensional structures, whose existence is made possible by the wave function's structure. These three-dimensional entities, though derivative, are thin in the sense that they have no inherent properties other than occupying a position in space. Thus, the wave function realist view is doubly structuralist: the wave function's structure allows three-dimensional entities to emerge derivatively, and it also governs how these entities behave. This stands in contrast to my view, where the wave function does not play a material role, and the fundamental entities exist directly in spacetime, with the wave function merely serving as part of the nomological architecture.

Between Properties and Structure

In the previous section, I presented a framework in which thin objects—identical, property-less entities—are embedded within a structuralist metaphysics. This contrasts with the traditional metaphysical view, which centers on objects defined by intrinsic properties. I explained how, in this alternative framework, all fundamental entities are indistinguishable in kind, and their behavior is governed (or characterized) by a common fundamental law. Despite this, the framework can still account for the wide variety of observable behaviors in the world. Crucially, this is achieved not by appealing to intrinsic properties, but by implementing effective laws that differentiate behavior based on context. In what follows, I argue that this alternative is not only viable but superior to both traditional object-oriented, property-based ontologies and other structuralist approaches. *My proposal offers a more economical and simpler metaphysics, which also provides greater explanatory power and aligns more naturally with the theoretical structure of quantum mechanics*—whether one adopts my framework or supports approaches in which the wave function represents matter. At the end of this section, I will address potential criticisms of this perspective, offering responses that clarify how this approach navigates and resolves the issues raised.

Advantages over the Traditional Metaphysics

A key motivation often cited in favor of the traditional view is its supposed simplicity: it offers the most direct route to accounting for the behavior of fundamental entities. But is it truly the simplest? Within that framework, much of the explanatory weight rests on the notion of properties. However, as Esfeld (2014a) has emphasized, and as I have briefly touched upon, our current understanding of both categorical and dispositional *properties remains highly problematic*. In contrast, the framework I propose bypasses this entire debate by eliminating the need for fundamental properties altogether, aside from spatiotemporal characteristics.

Furthermore, my account is *ontologically simpler and more parsimonious* than the traditional alternative. The standard view operates with three metaphysical categories beyond spacetime: the ontology, its fundamental properties, and the laws of nature. My approach, by contrast, requires only two: the ontology and the laws of nature (broadly construed as nomological structure). By positing fewer ontological kinds, my view aligns more favorably with Ockham's razor. This principle, often invoked in theory choice, suggests that, all else being equal, we should prefer the explanation that makes the fewest assumptions. While notoriously difficult to defend rigorously (see Sober 2015, Baker 2003, 2007), parsimony has often guided successful theory development. Indeed, one might even craft an optimistic meta-inductive argument in its favor: my account postulates just one kind of entity, and historically, theories that reduce ontological commitments in this way have often outperformed less parsimonious alternatives. Consider examples from ancient cosmology, pre-Darwinian biology, or early physiology: each posited additional kinds (the ether, an intelligent designer, the Élan Vital) to explain phenomena. Later developments (Newtonian gravity, Darwin's theory of natural selection, and modern physiology) eliminated these extra entities and produced more empirically adequate theories with leaner ontologies. By analogy, it seems plausible that the traditional view, which posits more categories than mine, may likewise be incorrect.

Humeans might argue that, from their perspective, which does not regard them as ontological commitments, my appeal to greater parsimony is misplaced. Nonetheless, my approach can be seen as a *natural extension of Humeanism*: it results in a "Better Best System." In the Humean view, laws are nothing more than axioms or theorems of the best systematization of the world's mosaic of facts. My account offers a superior balance between simplicity and explanatory strength. Importantly, introducing effective laws in addition to fundamental ones does not complicate the system compared to the traditional picture: the theorems remain the same, and so do the axioms; the only shift lies in reinterpreting parameters as constants. However, my approach offers the added benefit of positing only one kind of matter, instead of several. Moreover, my framework could accommodate a version of Humean supervenience when paired with my foundational commitment to spatiotemporal ontology. Recall that in my view, the wave function is not part of the material world but rather a component of the laws. Assuming symmetry constraints can uniquely determine the fundamental law among many theoretical possibilities, a Humean might claim that the wave function supervenes on the local arrangement of matter—the configuration of the thin objects—much like what Esfeld (2014a) calls "super-Humeanism" (see also Hall 2015). In a non-Humean context, where laws are treated differently, the argument for parsimony may carry less weight. Even so, it can still be used to motivate the preference for my framework. It's important to highlight that my view does not require a definitive stance on the metaphysical nature of laws. And even if laws remain mysterious, it is arguably more coherent to concentrate that mystery in one place, within the laws, than to spread it across both laws and intrinsic properties.

This perspective may also appeal to *empirically minded philosophers* and scientists. After all, when distinguishing a proton from an electron in experimental contexts, what we observe is not charge or mass directly, but rather a difference in behavior, such as a divergence in trajectory in a bubble chamber. What is directly observable is motion: positions changing over time. Mass, charge, and spin are inferred, never directly seen. From a standpoint that privileges what is observable, it seems natural to adopt an ontology centered on positions. This, then, is the ontology we should treat as real: localized, moving entities.

One might question whether parsimony alone is sufficient, or whether it compromises explanatory power. The worry would be that my model, while ontologically modest, may fail to explain as much. But I would argue that it actually has *greater explanatory power*, precisely because there is less to explain. For one, a key advantage of the thin-object, law-based approach is that it avoids metaphysical duplicates. Traditional property-based metaphysics often leads to the problem of entities that are qualitatively indistinguishable in every way, yet still treated as distinct due to the possession of different intrinsic properties. For example, if two entities have identical positions and trajectories but differ only in some hidden property like mass or charge, this leads to a kind of metaphysical overpopulation where there are entities that don't actually seem to contribute to the explanatory picture in any meaningful way. The thin-object view avoids this issue by not positing intrinsic properties at all. Instead, it focuses on the relational nomic structure defined by the laws of nature. In this framework, identity and difference among entities are not determined by hidden properties, but by their roles in the larger network of laws that govern them. This view is consistent with structuralist approaches, which argue that what matters metaphysically is not the intrinsic nature of entities, but their structural relationships within the system (see next section).

Another advantage of the thin-object approach is that it sidesteps what one can call the "*glue problem*" of the traditional metaphysical frameworks. This problem arises from the difficulty of explaining how intrinsic properties and the "bare substratum" of an object come together. For example, if a particle is thought to be a bare substance that carries properties like mass, we are left with the task of explaining how these properties bind together to form the particle. In contrast, the thin-object view does not face this problem because there are no intrinsic properties to bind together in the first place. Instead, identity and behavior are determined by the laws of nature and the relationships between objects. Since the law-based approach doesn't invoke the concept of an intrinsic substance "holding" properties, there's no need for a metaphysical glue to bind them.

Furthermore, this approach is able to explain *why certain properties emerge* while others do not. Traditional metaphysical frameworks struggle with the question of why certain intrinsic properties, such as mass, charge, or spin, are considered fundamental, while others (e.g., color, temperature) are treated as emergent. The thin-object view bypasses this issue by rejecting intrinsic properties altogether and thinking that all properties emerge. Rather than treating mass, charge, and spin as fundamental characteristics that particles "have," the view treats these features as emergent consequences of the laws that govern matter. This allows for a more coherent and parsimonious explanation of why certain properties are more fundamental than others: they emerge as part of the laws that dictate the behavior of matter. The effective laws that describe particles' behaviors give rise to these features, rather than the properties being inherently possessed by the particles themselves.

Moreover, in traditional accounts, one must explain *why intrinsic properties have the values they do*. Why do protons and electrons have the particular masses and charges that they do? This is a mystery in the traditional understanding. Instead, in the thin-object framework, this problem disappears because no intrinsic properties are posited to begin with. The world consists of thin objects whose existence and behavior are entirely determined by the laws that govern them. There's no need for a metaphysical explanation for that specific value rather than another one. Because of this, the thin-object approach provides a more parsimonious and explanatorily powerful framework. Someone might object that I still haven't explained why the effective laws are as they are. But this concern applies equally to the traditional view: we don't know why its laws are what they are, nor why particles possess the properties they do. In my framework, the burden shifts: the only thing requiring explanation is the structure of the laws. And as previously discussed, symmetry constraints and the principle of simplicity may offer a path to such explanations.

Be that as it may, my perspective aligns well with the theory of relativity, according to which *mass varies with velocity*. One might attempt to rescue the property-based view by suggesting that rest mass, the component that remains invariant with velocity, is the true property. But this leaves one with the burden of explaining what the additional, velocity-dependent portion of mass represents. On the other hand, if we abandon the notion of mass as a property altogether, such complications no longer arise.

Within classical electrodynamics, my framework also offers a coherent account of certain *asymmetries between particles and fields*. One striking asymmetry is that particles have properties, while fields do not. This raises a deeper question: Why are there various particle types with different charges and masses, but only a single electromagnetic field? My approach faces no such issue: it posits only one particle kind, with no intrinsic properties, and no need for fields. Additionally, the standard view holds that particles generate fields, which in turn act on particles through their properties, yet the reverse interaction, fields responding to particles through their own "properties," does not occur, as fields lack such attributes. These inconsistencies disappear entirely under my account, where fields are removed from the ontology, eliminating the need for any explanatory asymmetry. (A similar concern arises in discussions of the pilot-wave theory. A common objection is that the wave function appears to influence the particles without any reciprocal action: particles do not, in turn, influence the wave function. From the standpoint of my approach, this is not problematic: the wave function is understood as part of the law, not as a material entity, and laws govern without being affected by what they govern. Matter does not "act" on a law.)

Even more compelling is how my approach resolves the infamous issue of *contextuality* in quantum mechanics, previously introduced in Chapter 4 and that we already discussed in this chapter. The concept of spin has long been taken as a quintessential quantum property. Yet its contextual nature—its measured value depending on the particular setup of the measurement—poses a serious challenge. Imagine measuring the length of a square table and finding that the result changes depending on whether you measured it alongside its weight or its height. Such behavior calls into question whether spin can genuinely be considered a fundamental property. What kind of intrinsic feature behaves this way? Under my proposed framework, there is no such mystery: spin is not an intrinsic property, but part of the dynamical law that governs particle motion. The observed contextuality simply reflects the fact that the experimental setup interferes with the system, altering what one might otherwise interpret as a pre-existing value. In my approach, this interpretation is already well-supported: spin is integrated into the law, not as a property, but as a structural component of the dynamics, rendering its contextual nature entirely unproblematic.

To summarize, my thin-object proposal proves to be more adaptable than the traditional property-based view to the quantum domain. Moreover, it offers a simpler, cleaner, and more explanatory and coherent framework that avoids the conceptual pitfalls associated with traditional metaphysical categories.

Advantages over Structuralism

One of the first similarities between my view and structuralism is that both rely on the concept of structure, understood as a network of relations. In eliminativism, there are no definite *relata*, while in moderate structuralism, objects are defined in terms of their relations. In my view, objects are thin and defined by these relational structures. Both structuralism and my approach *align well with quantum theory*, in contrast to the traditional view. Structuralism, in part, arose as a way to understand quantum mechanics, and while my account does not arise from this motivation, it is still natural to extend the idea that quantum properties, like spin, are not intrinsic. Thus, in both the quantum and classical domains, one might conclude that what we call intrinsic properties, such as mass and charge, are better understood as part of the laws that describe interactions.

Additionally, both approaches have very specific ways of understanding what the *wave function* is, though we interpret it differently. French (2013) has noted that the wave function is underdetermined. However, this underdetermination can be resolved by understanding the wave function as being "constituted by the laws and the associated symmetry principles." This view posits that the wave function emerges from the symmetries, thus defining it in terms of the fundamental structure. Instead, I see the wave function as part of the definition of the law. In both structuralism and my approach, the wave function is not treated as a material object, nor is it necessarily "real," unless one adopts a realist view of laws, which is not required in either framework. Both views are compatible with Humeanism, where laws are simply the axioms or posits of our best theories, although primitivism or necessitarianism is not excluded *a priori*.

Both frameworks also employ *simplicity and parsimony* as guides to metaphysics. Structuralism eliminates objects in favor of structure, while I eliminate properties. However, both share a skepticism toward properties and a corresponding inclination to have laws and symmetries do the modal and explanatory work. Moderate structuralism differs in that it does not entirely discard properties; instead, structure is made up of relational properties and intrinsic properties (Esfeld and Lam 2011).

A final commonality between my approach and structuralism is the significance of *symmetries*, although the treatment of them differs. In my framework, the symmetries of the theory help define physical objects and constrain the form of the laws. Symmetries, together with thin objects, generate the structure, which includes both the laws and the properties. In contrast, structuralism sees symmetries as ontologically prior to objects, such that symmetries and structure together generate laws, objects, and properties.

This brings us to the main difference between the two views: *my approach remains object-oriented*, though the objects in my framework are thin. In eliminativism, structure through symmetries grounds thick objects and their properties. In my view, thin objects are grounded in structural relations between them through symmetries. Thus, in this respect, my approach aligns more closely with traditional object-oriented metaphysics than with eliminativism. Despite both being object-oriented, my view differs from moderate structuralism. In moderate structuralism, objects and structure are ontologically on the same level: structure is a network of relational properties between objects, which are defined in terms of them. Objects may have intrinsic properties like mass or charge, but they are primarily defined by their relations, which represent the ways they could be. In my view, this is not the case: I posit only spatiotemporal relations, but not intrinsic or relational properties that account for motion (nomological relations). Thin objects, in my approach, are interconnected by laws, not properties. Laws do not define objects, and objects do not define laws.

In conclusion, my approach is consistent with structuralism in that we have thin objects without intrinsic properties and structures, conceived as the nomological network required to specify the motion of objects. However, it is not an eliminativist view, as it still acknowledges the existence of objects. Nor is it a variety of moderate structuralism, since objects and laws are not ontologically on the same level. Objects constitute matter, while laws either supervene (if Humeanism holds) or not on them, but do not define the objects.

My framework can also account for both *underdetermination and entanglement.* Regarding underdetermination, for example, in quantum mechanics, the evolution of the wave function over time is not crucial, as long as the law governing the wave function for matter remains the same. Different wave functions may produce identical behaviors for objects, so the evolution of the wave function is underdetermined. However, this underdetermination is not problematic, as the wave function is not material. Rather, underdetermination concerning objects is resolved by positing fundamental objects as the simplest and most unifying explanation, constrained by symmetries. This view also provides a useful perspective on entanglement. While subsystems in an entangled state lack their own wave function, this does not imply that the individuals represented by the subsystems cease to exist. The wave function does not describe matter; rather, thin objects do, and the dependence captured by entanglement is explained in terms of laws.

As noted in Chapter 2, eliminativism faces the intelligibility problem: it makes no sense to say relations exist without *relata*. Moderate structuralism introduces *relata* to avoid this issue, but it immediately faces the problem of properties. This seems to contradict one of the core motivations for structuralism, which sought to eliminate properties. My approach *sidesteps the intelligibility issue* by maintaining that there are *relata*—specifically, thin objects. Additionally, it avoids the problem of properties in moderate structuralism, as it replaces properties and laws with laws alone.

In conclusion, I have argued that my approach shares many motivations with structuralism, but it addresses fewer objections. Therefore, if one is already persuaded by the structuralists' arguments that radical changes are needed, my view offers a less problematic alternative.

Anticipating and Answering Objections

One potential objection to this view is that it appears overly *radical and revisionary*: Why would anyone want to discard properties when they work so well within the traditional framework? In response, I would argue that the proposal is not unnecessary. Properties are notoriously difficult to define, and, as previously discussed, there are significant challenges when attempting to specify what fundamental properties actually are. Furthermore, as outlined in the previous section, my approach offers several advantages over the traditional model, making it worth exploring: it aligns with Humean supervenience, adapts well to the quantum domain, and is simpler and more explanatory. An objection may arise regarding the *consistency* of primitive postulates. One might complain that not all primitive postulates are on the same footing and that mine are too radical. If that's the case, we must be relying on different notions of what counts as a radical postulation. In my opinion, the use of properties to explain particle behavior in the traditional view seems almost magical. What exactly is mass? Where is it located? How does the law interact with particle mass to modify its behavior? In my approach, these questions don't arise, except for one: How does the law interact with the particle to alter its behavior?

Similarly, one might argue that asserting the nonexistence of properties *contradicts our intuitions and ordinary beliefs*. This perspective suggests that such a view would render many of our common beliefs false, and that theories which do so are generally not preferred, all else being equal. However, one can reject this criterion or argue that not all things are equal. Many modern scientific theories ask us to revise our ordinary beliefs, but we don't dismiss them solely for that reason. These theories provide explanatory power and are more empirically adequate than our everyday beliefs. For instance, we once believed that matter was continuous, but atomic theory, which is more explanatory and empirically successful, revealed that matter is mostly empty space. Similarly, our belief in the existence of properties is mistaken, as my approach, which offers a more explanatory and empirically robust framework, shows that properties do not exist.

Another concern might be the *apparent mystery* of how different effective laws act on matter if there is only one kind of matter. Traditionally, it is believed that the positive charge of a proton determines its motion in a magnetic field. Under my proposed view, however, we have only matter. So, how is a material entity paired with its corresponding effective law? Is this not just another "*glue problem*," similar to the issue of how particles are paired with their properties in the traditional framework? In response, it's important to note that in the traditional view, not only is it mysterious why a particle is paired with its properties, but it's also a brute fact that the positive charge results in a certain behavior in a magnetic field. In contrast, under my view, it is a fundamental fact that effective laws govern the behavior of material entities. Every view requires primitives, and in the traditional approach, fundamental properties are taken as primitive. In my view, the primitive fact is the pairing between objects and their effective laws, or more specifically, between thin objects and the structure. Instead of focusing on laws, it is more productive to think of them as part of a structural network that governs how matter moves.

A further question may be raised about *what differentiates empty points from material points* in my framework. The traditional response is that material points are simply points "dressed" with mass, but I find this unsatisfactory, especially considering the mysterious nature of mass itself. Nevertheless, it's not accurate to say I have nothing. In my view, there is a fundamental distinction between space points: some are "filled with matter," while others are not. The material points change their location over time according to the laws.

There may also be more technical issues to consider. For example, one might argue that the equation $E = mc^2$, which *connects energy to mass*, cannot be reconciled with my approach, since if mass is not a property, neither is energy. However, there is no intrinsic problem with this, and no catastrophic consequences arise if energy is not regarded as a property. While we are accustomed to thinking of mass as a property, this view is mistaken. Similarly, we are also accustomed to thinking of energy as a property, but that belief is similarly incorrect.

A *tu quoque* objection might also be raised, suggesting that my approach is subject to the same critique I made of wave function realism in Chapter 3, which used principles to explain the emergence of the world as we see it. One could argue that my approach is similarly flawed because all properties, both intrinsic and macroscopic, emerge. In reply, I would clarify that wave function realism is problematic because it posits that spatiotemporal ontology itself emerges through principles, and without a fundamental spatiotemporal ontology, no constructive explanation is possible. Once three-dimensional space emerges, wave function realism can proceed constructively. However, this is not the case in my approach, where I start with a fundamental spatiotemporal ontology and allow properties to emerge dynamically, not *via* principles.

An objection might be raised claiming my approach is *conspiratorial*, thus implausible, following something similar to the no miracle argument for realism. According to this argument, the empirical and explanatory success of a theory can either be attributed to its truth or to an implausible conspiracy in which everything behaves as if the theory were true, even though it is not. Thus, one might suggest that my approach is too miraculous to be plausible: that there are no intrinsic properties, yet everything behaves as if they exist. In response, I would argue that this charge is exaggerated. In my view, properties are not fundamental, but they do emerge dynamically. Since properties exist, the analogy to a conspiracy is flawed.

Lastly, there may be *misunderstandings* regarding the nature of my approach. One might believe that my view entails infinitely many laws, one law for every massive object, suggesting that my theory is not more parsimonious than traditional views. However, this is not the case. Rather, in my approach, there is one type of material entity and one kind of law. The law acts differently depending on what it governs. If we think structurally, the law and its effective forms represent the network of relations between the fundamental objects. Macroscopic objects are composed of microscopic entities, and their behavior is explained in terms of these entities. In the traditional view, we might have N fundamental distinct particles with their properties and one law of nature. In my framework, there is only one type of particle, but N effective laws, or more precisely, one type of particle and a more complex structure.

Conclusive Remarks

In this chapter, I have outlined an alternative to the traditional approach used by naturalized metaphysicians to explain the behavior of physical objects. While the conventional view relies on laws of nature to guide objects' motion, with properties such as mass or charge playing a crucial explanatory role, my proposal presents a different perspective. In my view, fundamental objects lack any intrinsic properties other than those necessary to define their nature. In the traditional perspective, particles are categorized based on their distinct properties. In contrast, in my account, all matter is identical at a fundamental level. The reason why two objects might behave differently in the same situation is not due to their differing properties but because they obey different effective laws—variants of the same fundamental law that encapsulate what we typically call properties. For example, whereas the traditional view explains the upward motion of an electron and the downward motion of a proton in a magnetic field by referencing their opposite charges, I propose that there is no fundamental difference between an electron and a proton. Both are particles, but they are governed by different effective laws.

This *thin-object-oriented metaphysics, grounded in structure,* stands in contrast to the traditional *thick-object-oriented metaphysics, which is rooted in properties.* In my framework, objects are "bare" or "thin," lacking fundamental properties. The view is *structuralist* because the effective laws create structural relations between the fundamental objects. Symmetries also play a significant role, as they help determine both the ontology and the laws of nature. I have argued that my approach is *more explanatory and simpler* than the traditional one and that it is *consistent with quantum theory and Humean supervenience.* Specifically, I contend that, within the quantum domain, this approach provides the best metaphysical framework, whether one subscribes to my spatiotemporal foundational approach or wave function realism.

That said, my approach is highly speculative. Is it truly desirable to place everything within laws and define laws structurally? I recognize that this could be viewed as excessive, and I wouldn't be surprised if, ultimately, more suitable alternatives emerge. Nevertheless, given the current state of the discussion, I believe it is worth further exploration.

Further Readings and Discussion

The literature on the topic discussed in this chapter is extensive, covering much more than what has been addressed here. Informative introductory works on properties include Choi and Fara (2021), Allen (2016, 2018), Marshall and Weatherson (2018), Dorr (2022), O'Connor (2021), and Orilia and Paolini Paoletti (2022), along with Bird and Tobin (2023).

For discussions on laws, see Hildebrand (2020, 2023), Bhogal (2020), Carroll (2022), and Swartz (2022). A comprehensive treatment of the relationism-substantivalism debate in classical spacetime is provided by Hoefer, Huggett, and Read (2023).

On fundamentality and related concepts, as they pertain to the issues explored in this book, see Tahko (2018), Tahko and Lowe (2020), Bliss and Trogdon (2021), and Rickles (2022).

7

"What If?"

This chapter offers a reflection on the *significance of the historical sequence* in which certain events unfolded, as well as on how individuals in positions of power leveraged their charisma and privilege to promote particular viewpoints while suppressing or sidelining others. The argument is that these historical and sociological factors played a substantial role in shaping the widespread belief that the wave function should be regarded as a real physical field—or at least as an indispensable component of quantum theory—prompting many to feel compelled to radically revise their entire scientific worldview. However, *these motivations are ultimately unwarranted.* Had history taken a different course, and had certain figures not wielded such influence, an approach more in line with the one I am proposing—one that preserves greater continuity with classical traditions and does not demand a complete theoretical overhaul—would likely have been received more favorably and perhaps even readily embraced.

Chapter Schema Details

As discussed in the first two chapters, it is widely held among realists that for quantum theories to align with scientific realism, they must resolve the measurement problem. It is also common for realists to view these theories primarily as accounts of the wave function.

In this chapter, I challenge both assumptions. I argue that *the centrality of the measurement problem in realist discourse is largely a historical accident*—shaped by misinterpretations of theoretical and experimental developments, a particular philosophical climate, the influence of charismatic figures, and other nonscientific factors. Classical physics offered constructive explanations of macroscopic phenomena by appealing to the motion of microscopic constituents. Since quantum mechanics initially failed to provide such a constructive framework, it would have been natural to seek a deeper understanding, as had been the norm in similar contexts. In this chapter, *I explore how things might have turned out if everyone had agreed that the right direction was to look for a constructive explanation.* Realists would have approached quantum theory as they did thermodynamics: as a principle theory in need of a constructive underpinning. Just as statistical mechanics explained thermodynamic laws, quantum mechanics would have been seen as requiring a similar explanation. Under this alternative path, *realists would not have been focusing on the measurement problem*. There would have been no turn toward ontologies in high-dimensional configuration space, no appeal to undetectable many-worlds, consciousness-based collapse theories, or stochastic nonlinear modifications of the Schrödinger equation. Instead, attention would have shifted on developing a theory in which particles evolve deterministically according to nonclassical and nonlocal laws—something akin to the pilot-wave theory.

Then, I return to the *actual historical trajectory*, arguing that the measurement problem was introduced as *a strategic way to engage instrumentalists*—who were largely indifferent to realism but held greater influence—by framing the issue in terms of empirical adequacy. I conclude that *resolving the measurement problem is insufficient for developing a satisfying realist interpretation of quantum theory* in a constructive way. At most, it highlights an empirical shortcoming. What is needed is a theory that constructively explains the phenomena, not merely one that patches over their paradoxes.

The remaining challenge is how to understand nonlocality, as revealed by the violation of Bell's inequality. In this context, at the end of the chapter I discuss two strategies for avoiding the conclusion of nonlocality—*superdeterminism and retrocausality*—but I argue that their implications seem even more radical than accepting nonlocality itself.

An Ideal Quantum Framework

As discussed throughout this book, quantum theory is often presented as requiring the *abandonment of the idea that phenomena can be constructively explained*, as was customary in the classical domain. This was the position, in various forms, of the members of the Copenhagen school. They urged us to accept that only what is directly observable can be meaningfully discussed, and that any attempt to explain the quantum world in the same constructive manner as classical physics was misguided or even meaningless. Over the years, this message has been ingrained in generations of students. Yet, as we have seen with the alternatives discussed so far, this is not a necessary conclusion of quantum theory—*it is a Dogma*, what Beller (2001) aptly described as the "rhetoric of inevitability" promoted by Bohr. In reality, this view owes much of its persistence to the sheer effectiveness of the propaganda surrounding it. This narrative has been scrutinized in existing literature—most notably by Cushing (1996) and Becker (2018)—but here I intend to go a step further. Rather than merely critiquing the supposed inevitability of the Copenhagen interpretation, I aim to invert that claim: *what should have been seen as inevitable was the realization that the natural continuation of the constructive explanatory framework led directly to the pilot-wave theory.* Instead, this option has received disproportionate and unjustified skepticism.

Let us consider how things might have progressed had the Copenhagen dogma not prevailed. In the following section, I present *a series of arguments* offered at the time—not to focus on how they were misunderstood or misrepresented, but *to trace where they ought to have led.* And that destination, as I argue, is quite different from where we ended up. If I may end on a lighter note, this thought experiment shows that our world is not, in fact, the best of all possible quantum worlds.

A Principle Theory for Quantum Phenomena

Let us go back to the history of quantum mechanics and start from the beginning, to construct *a fictional story* of how things could have played out. As briefly sketched in Chapter 2, quantum theory began as a *minor adjustment to the classical picture* in order to explain experimental results—such as the blackbody radiation problem, atomic spectra, and the stability of the atom—that did not fit within the classical schema. The data suggested that certain quantities were not continuous, but quantized, and the goal at the time was to develop a principle theory that would establish the correct quantization rules for various quantities. This was done using the *Bohr-Sommerfeld quantization conditions* within the classical framework that viewed matter as particles and light as a wave. However, constructing a coherent understanding proved difficult, as some experiments contradicted the wave nature of light (such as the Compton and photoelectric effects), while others questioned the particle nature of matter (e.g., particle interference). Nevertheless, evidence began to emerge indicating that *both particles and waves* might be needed. For instance, in 1905, Einstein proposed that light was composed of particles, or photons, which could also explain the Compton effect. De Broglie, in 1924, expanded on this idea, suggesting that photons were associated with light waves (rather than being identical to them), with the light wave "guiding" the photons' motion. He also proposed the existence of "matter waves" associated with particles. However, de Broglie had yet to provide a viable dynamical equation for these waves at that time.

An Attempt at the Classical Schema with Wave Mechanics

In a series of papers written in 1926–7, Schrödinger began to develop a wave-only constructive theory, known as *wave mechanics*, with the aim of *constructively explaining* the origin of the quantization rules. According to this theory, the fundamental object is a wave (in the spirit of de Broglie's matter waves, but without involving particles), which evolves according to the equation now associated with his name. Schrödinger's wave, when written as a function of position, is now called the *wave function*. The stated goal of wave mechanics was to provide a *visualizable* depiction of what occurred inside an atom, similar to the explanatory constructive framework that had been successful in the past. Schrödinger demonstrated that the hydrogen spectrum could be explained through the nodes of this wave. Particles, in this context, are seen as *narrow wave packets* that, when observed from a distance, give the impression of being point-like particles.

Even if this approach looked very promising, it eventually became clear that there were serious *shortcomings*. One major issue was that *wave packets spread* rapidly, which made it impossible to account for the stable "particle" trajectories that were

a theory from falsification, if one allows using far-fetched ideas. For instance, perhaps aliens are hypnotizing us into believing that there are no macroscopic superpositions. Or one could think that there are unobservable parallel worlds in which each term of the superposition is realized, but we do not see them because the various worlds do not interact. Or we can imagine that conscious observers interact with matter in such a way as to destroy macroscopic superpositions. Or one can consider the possibility that logic is false, and that logical contradiction, like a cat is dead and not-dead at the same time, is not a problem. Nonetheless, we do not seem to have independent grounds to believe any of these possibilities: we have no independent evidence of massive deception, undetectable worlds, the nonclosure of the physical world, or the invalidity of the principle of noncontradiction. Thus, the most straightforward, plausible, if not rational, conclusion to draw from the unobserved macroscopic superpositions is that *a new ontology for quantum theory might be advisable*. At the very least, it would be worth exploring the consequences that such a change in ontology would produce.

The Constructive Quantum Schema of the de Broglie-Bohm Pilot-Wave Theory

A proposed solution to these problems arrived in 1952 with Bohm, who built on de Broglie's idea of having both particles and waves. However, he proposed that *particles constitute matter*, while the wave does not. Rather, the wave "guides" the motion of the particle, hence the name of *pilot-wave theory*. Bohm started from Newton's equation $x, y, z.$ and supplemented it with *a quantum potential* term, $Q = Q^{\psi}$, in which a Schrödinger-evolving wave function appears: $ma = F + Q^{\psi}$. In this theory, the wave function is a convenient tool to describe the motion of the particles through the quantum potential, which acts similarly to gravitational or electromagnetic potentials, as it affects the trajectories of the particles.

Assuming that particles are the fundamental constituents of matter and that the wave function is a tool for describing their interaction, the problem of spreading wave packets disappears: wave packets do not represent the fundamental nature of what we used to call "particles," so whether they spread or not is irrelevant. Moreover, the superpositions of the wave function characterize the strength and modality of the interaction—not the status of objects—so *the cat is never in superposition* because the cat is made of particles.

Here are some features of the theory, in comparison with the classical framework. Since Bohm's equation for the particle stems from a modification of Newton's second law, it is deterministic. Thus, one might expect that experimental outcomes could be predicted precisely. However, this is not the case. In fact, it turns out that it is impossible to know the particles' locations with absolute precision. Not knowing the initial conditions—even with deterministic laws—leads to *probabilistic predictions*. That a deterministic theory produces probabilistic predictions is not too surprising, as this also happens in statistical mechanics. If we believe Newtonian mechanics is universal, then we should be able to predict how large systems, such as gases, behave. However, since such systems are large, it is difficult to solve all the equations, and our information about the initial conditions of all the particles is limited. Thus, we cannot arrive at precise laws. Instead, statistical methods are used, leading to probabilistic macroscopic laws such as the second law of thermodynamics, according to which entropy increases probabilistically. One may wonder why we cannot know the positions of particles with arbitrary precision. Where does this *absolute uncertainty* come from? Again, statistical mechanics offers a helpful analogy. When we want to extract information from a system, we must interact with it. For example, to measure my temperature, I use a thermometer. The thermometer interacts with my body; our temperatures change until equilibrium is reached, and the thermometer displays a number. This is the equilibrium temperature, which, due to how the thermometer is designed, is very close to my initial temperature—close enough that I can take it as such. In this way, I've gained information about one system (myself). However, if two systems are in equilibrium, no more information can be extracted. Something similar happens here: the particles are in "*quantum*" *equilibrium* with respect to the wave function (which is not in equilibrium): the distribution of their position is given by the squared modulus of the wave function. So, the maximal information we can extract from any system about where the particles are is given by their wave function. This limitation of what we can know about a system was first discovered by Heisenberg in 1927, in the form of the *uncertainty principle*: the more precisely we determine a particle's position, the less we can know about its momentum (though Heisenberg took his principle in a much stronger sense, as discussed in Chapter 2 and later). To summarize: in this theory, we can only predict how outcomes will be probabilistically distributed. When performing an experiment to determine where a particle is, the probability P of finding it at a given position x is provided by the square modulus of the wave function evaluated at that point: $P(x) = |\psi(x)|^2$. This is the *Born rule*, proposed in 1926. To draw an analogy between statistical mechanics and the pilot-wave theory: while in statistical mechanics a system likely evolves toward higher entropy, here the system likely evolves toward regions where the square modulus of the wave function is greater.

As in the classical framework, all the properties we attribute to systems are ultimately functions of the positions of particles, and all experiments ultimately yield results that describe the location of something. For example, the velocity of a system is a function of its position, and an experiment designed to measure a particle's velocity is essentially measuring how far the particle has moved over a given time period. Similarly, the weight of an object is indicated by the position of the scale's pointer on a specific value, and so on. Generally, therefore, *all measurements can be considered position measurements* (Bell 1982). Nonetheless, as discovered by Heisenberg, Born, and Jordan in 1926, it is possible to effectively summarize experimental results in terms of matrices or linear operators. For convenience, one can associate a suitable *linear operator* $\hat{A}$ to any experiment, such that the possible experimental results are the eigenvalues a_i of that operator: $\hat{A}\,\psi_i = a_i\psi_i$. If one writes the wave function ψ as a linear combination of the corresponding eigenstates, $\psi = c_1\psi_1 + \cdots + c_i\psi_i + \cdots$, the probability P of obtaining a particular result a_i is given by the square of the coefficient c_i with which the eigenstate ψ_i appears in the combination: $P(a_i) = |c_i|^2$.

This is the generalized Born rule, which gives the probability of finding a given experimental result.

In classical physics, it is safe to assume that *measurements reveal properties* of a system prior to measurement, as in the thermometer example above. That is, it is safe to assume that the system–apparatus interaction can be neglected, and instruments are designed specifically to interfere as little as possible with the object being measured. This means that in classical mechanics one can have *genuine* measurements. However, as Bohr (1928) and Heisenberg (1927) emphasized, this is no longer feasible when examining small systems. For example, to determine the position of a stationary particle, we must interact with it—such as by illuminating it with light—which changes its position. The more accurately we try to measure its position, the more we disturb its velocity, making it harder to measure, as the uncertainty principle dictates. In this sense, the experiment is destructive. In 1932, von Neumann showed that *most experiments are indeed destructive* (though, as noted, this is not what he himself took to have proven—more on that later, and in Chapter 4).

Sometimes, these impossibility proofs of genuine measurements are referred to as *contextuality* proofs, to emphasize that whether an experiment is destructive depends on the measurement context. A series of destructive experiments (which, therefore, cannot be considered to be measuring anything) are associated with operators that *do not commute*—that is, the value of their product depends on the order in which the operations are performed. This makes intuitive sense: *when an experiment is destructive, the order matters*. If you read a message and then burn it, it's quite different from burning it and then reading it.

Empirical Evidence of the Constructive Quantum Theory

As discussed in the previous section, the predictions of the pilot-wave theory are given by the Born rule. The theory was *immediately confirmed*, accounting for all the experiments that initially challenged the quantum framework, such as atomic spectra, the photoelectric effect, the Compton effect, and so on. In particular, the pilot-wave theory successfully explained the interference and diffraction patterns in the two-slit experiment by providing the empirically correct probability distribution for the particles' positions. Moreover, it offered many more predictions that were later verified across *a wide range of phenomena*, such as tunneling effects. The highly nonclassical nature of particle trajectories was first observed experimentally in a 2011 two-slit experiment (Kocsis et al. 2011), where the *actual trajectories were measured* point by point. They revealed a behavior consistent with the theoretical trajectories computed by Philippidis, Dewdney, and Hiley (1979). (To clarify, it was difficult to measure these trajectories, which is why the experiment mentioned above took a long time to realize. Indeed, one might have thought it impossible, given the uncertainty principle. The technique used in this case relies on the notion of *weak measurement* [Aharonov, Albert, and Vaidman 1988], combined with an ingenious experimental procedure. A weak measurement is a genuine measurement in the sense that it does not significantly disturb the system. However, because of the uncertainty principle, the result it yields is highly imprecise. Suppose one performs a weak measurement of the particle's position x_1 at time t_1. The outcome will be a rough, imprecise estimate of x_1. By repeating the same weak measurement many times at the same position, one obtains a statistical distribution of values for x_1. Taking the average of these results provides a reliable estimate of the particle's position at time t_1. Then, at a later time t_2, a precise measurement of the particle's position x_2 is performed. From this, one can determine the particle's velocity at x_1 by calculating the difference between the two positions x_2 and x_1, divided by the time interval $t_2 - t_1$. By repeating this entire procedure, one constructs a velocity field, from which the particle trajectories are derived by taking the tangents to this field.)

The Surprise of Nonlocality

The pilot-wave theory appears to be a *very natural extension of the classical schema*: particles move, and even if we cannot predict exactly where they will end up, we can still obtain probabilistic predictions. Nonetheless, there is a troubling aspect—namely, *the interaction is explicitly nonlocal*: there are instantaneous influences across arbitrary distances. This stands in contrast to the idea that interactions propagate at finite speed, and to relativity, where an upper limit to any velocity is imposed. While it's true that even the classical notion of force is nonlocal, that kind of nonlocality can be eliminated. For instance, electromagnetism can be reformulated as a local theory by replacing forces with electromagnetic fields, which propagate at the speed of light. Likewise, it is possible to remove the nonlocal character of gravity by modifying the geometry of spacetime such that gravitational force is no longer required. In the pilot-wave theory, however, nonlocality is built directly into the definition of the wave function, which is formulated as a function of all particle configurations simultaneously—so it is truly explicit. Although it was unclear how to eliminate this feature, it was evident that doing so was necessary. In fact, *nonlocality seemed unacceptable* for the reasons we already saw: it is counterintuitive on many levels, and it is in contrast with the relativistic requirement that there is a maximum velocity.

Accordingly, in 1964, Bell set out to determine whether an alternative kind of constructive explanation for quantum phenomena could be developed *under the assumption that locality holds*. He developed a theoretical framework involving Schrödinger-evolving wave function governing the behavior of other, *hidden* or currently unknown, variables, and then he derived the empirical predictions of such a model. These predictions differed from those of the pilot-wave theory, thus opening the door to a *crucial experimental test* (Bell 1964). The first attempt to carry out this test was made by Freedman and Clauser in 1972, using an inequality—dubbed CHSH (Clauser et al. 1969)—which was a variation of Bell's original formulation. Later experiments, most notably those by Aspect and collaborators in 1982, produced *results consistent with the predictions of the pilot-wave theory*. As a consequence, *the assumption of locality had to be abandoned*, and the pilot-wave theory came to be seen as the most viable constructive framework for explaining quantum phenomena. This outcome prompted a renewed focus on the implications for relativistic extensions of the theory, given that the principle of locality lies at the heart of the conceptual foundation of relativity.

This is where we stand today—or rather, *where we would stand if we existed in an ideal quantum world*, a world in which, I would say, there were no misunderstandings, no power plays, only sound or unsound arguments. A schematic drawing of the ideal timeline is shown is Figure 7.1.

In such a world, there would be no talk about wave-particle duality, complex, high-dimensional field ontologies; no bizarre macroscopic superpositions; no need for stochastic or nonlinear modifications to the Schrödinger equation; no mention of unobservable worlds; no contextual properties; and no appeals to consciousness. *All of these speculative ideas would have been discarded* the moment physicists realized that they could continue to understand the new phenomena using the same constructive strategies that had served them in the past. These strategies require making physics about a spatiotemporal microscopic ontology, and understanding macroscopic behavior as emerging from microscopic dynamics. There was simply *no compelling reason to develop anything other than a constructive quantum theory*. When scientists encountered the pilot-wave theory, it offered a coherent and straightforward constructive framework—a natural extension of the classical schema—as it is a theory in which particles evolve in space through time, albeit according to nonclassical laws. In this framework, *the revolutionary feature of nonlocality*—the true novelty of quantum phenomena—is made immediately evident. And yet, this idealized scenario is not the one that actually unfolded. The real history of quantum theory diverged significantly from this clean and coherent narrative. In the next section, I return to that history in greater detail than I did in Chapter 2.

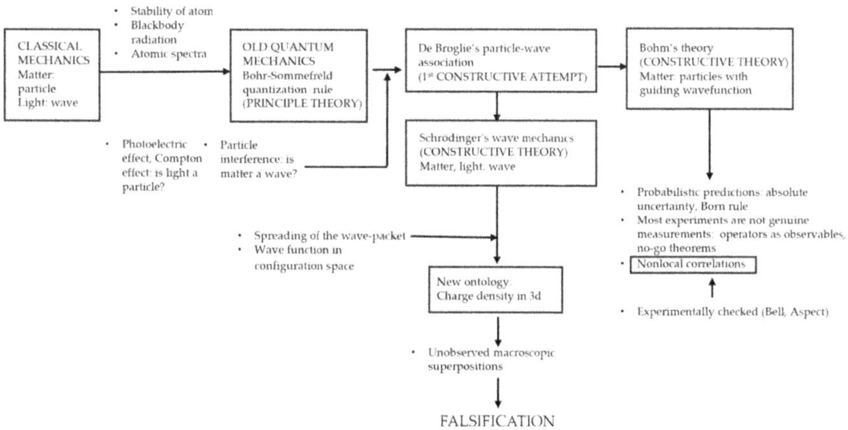

Figure 7.1 The "perfect" quantum timeline.

Going Back to Reality

Many books have been written to recount the development of quantum theory—most notably van der Waerden (1968), Jammer (1974), Fine (1986), Cushing (1996), Howard (1990, 2004, 2007), Freire (2015), Bricmont (2016), Norsen (2017), and Becker (2018).

As depicted in Figure 7.2, it is *not a linear story*, and it is very different from the tale we told in the previous section. It is rife with personal biases, chance events, poorly written papers, misunderstandings, power struggles, charisma—with individuals talking past one another, overlooking or misinterpreting arguments, and dismissing objections without valid justification. Here is a short overview. As discussed in the previous section, this would never have happened if people had not focused so heavily on the measurement problem.

1900–27: Matrices and Waves

As mentioned in Chapter 2, the so-called "*old quantum theory*" operated by taking classical mechanics as its foundation and introducing quantization through arbitrary rules. It wasn't until 1925 that Heisenberg, Born, and Jordan formulated the first comprehensive and coherent framework to organize experimental data, relying on *matrices* or self-adjoint operators. This method stood in stark contrast to the traditional conception of what a fundamental physical theory should be. While classical mechanics described the trajectories of particles in motion, matrix mechanics abandoned that framework altogether: it did not offer a mechanics in the conventional sense. Instead, it posited that *self-adjoint operators correspond to measurable physical quantities*, aiming to characterize the possible outcomes of measurements. It made no claims about the nature of the entities possessing these properties, their locations, or how they evolve through space and time. Though formal and abstract, this method succeeded in reproducing the known quantization rules by using matrices, rather than

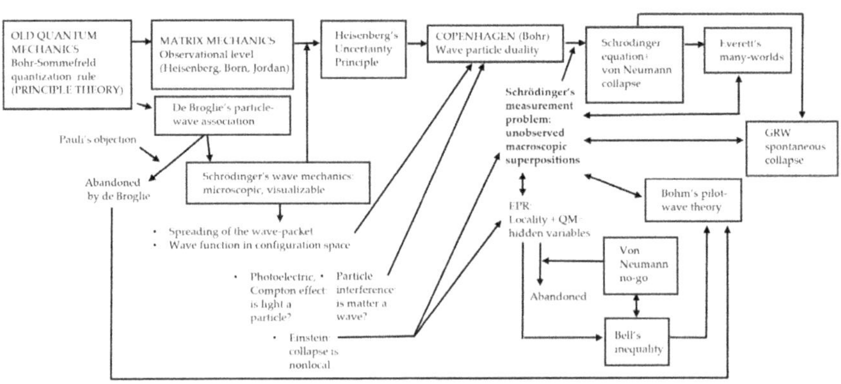

Figure 7.2 The actual quantum timeline (simplified).

deriving them from any underlying microscopic dynamics. This deliberate avoidance of reference to unobservable entities, along with the emphasis on observable outcomes, was clearly shaped by the rising influence of positivism at the time. Advocates of this framework came to believe that the persistent difficulty in producing a constructive account of experimental results was not due to the complexity of the phenomena or a lack of imagination, but because *the very idea of constructive explanation was flawed*. In their view, one should not attempt to understand quantum phenomena by describing the motions of unobservable entities but rather by offering a mathematically unified and aesthetically coherent formalism that exclusively refers to what is empirically accessible.

In the same year, Schrödinger challenged their position by introducing *wave mechanics*, which explicitly sought to offer a *constructive explanation* of quantum phenomena through a wave-based ontology. He also demonstrated that his formulation was mathematically equivalent to matrix mechanics, but contended that wave mechanics was superior in terms of explanatory power because it offered greater visualizability. In response, various scientists had different attitudes. On the one hand, Heisenberg moved away from the positivist defense of matrix mechanics and embraced Schrödinger's visualizability challenge. He adopted a particle ontology and, in 1927, argued that although particles do exist, they cannot follow definite trajectories because they are unable to possess well-defined positions and velocities simultaneously. He maintained that this conclusion was necessitated by the *uncertainty principle*, although, in truth, an alternative interpretation could have suggested that particles do have both properties, but we simply cannot know them with perfect accuracy.

On the other hand, Bohr strongly favored a wave ontology, which was more compatible with the atomic model he proposed in 1913. Therefore, he strongly opposed Heisenberg's particle ontology. Eventually, in 1927, he introduced his idea of *wave-particle complementarity* as a middle ground between these conflicting views (Bohr 1928). Born, in 1926, supported Schrödinger's wave interpretation, leading him to his statistical view of the wave function—that it represents the probability of detecting a particle during measurement. However, he later aligned himself with Heisenberg's perspective.

In any case, despite their internal disagreements, the core figures of the Copenhagen school, led by Bohr, presented a united front against Schrödinger's wave-based theory. Schrödinger's theory was accused of being inadequate for allegedly failing to account for quantum jumps. However, this was false: he had, in fact, published a paper explaining phenomena like the Compton effect and the photoelectric effect using wave vibrations.

Aside from the physicists around the Copenhagen group, during the 1927 Solvay Conference, the young physicist de Broglie presented an *early version of the pilot-wave theory*, a comprehensive quantum theory combining particles and waves. However, his idea was met with sharp criticism from Pauli, which likely played a significant role in de Broglie's decision to abandon the development of his theory for many years (as discussed in Bacciagaluppi and Valentini 2009).

1926–7: Unphysical Field Objection

The matrix and wave mechanics are an example of the underdetermination of theory by data: they are empirically equivalent but metaphysically incompatible. Arguing which is theoretically or explanatorily better was not really on the table, given the *positivistic climate* at the time, as such a discussion would have had to focus on super-empirical virtues. Still, there was no denying that Schrödinger's formulation proved to be more practical and user-friendly than matrix mechanics. As a result, it became the standard tool used by the physics community, even though most physicists adopted it without seriously engaging with its deeper ontological implications. Schrödinger, much like de Broglie before him and Bohm after him, found himself isolated in his efforts to explore the foundational meaning of his theory, eventually abandoning this line of inquiry altogether.

This outcome was largely driven by the tremendous influence, authority, and personal magnetism of Bohr, in contrast to others who lacked such advantages, and was further reinforced by the prevailing philosophical atmosphere steeped in positivism. Figures such as Einstein and Schrödinger notably pushed back against the Copenhagen stance, which insisted that the quantum realm must be accepted as fundamentally unintelligible. Despite their efforts, their *critiques were either dismissed outright or misrepresented.*

As discussed earlier, both Einstein and Schrödinger advocated for the idea that quantum mechanics, like thermodynamics, should eventually be grounded in a deeper, more complete theory, analogous to how statistical mechanics underlies thermodynamic behavior. The first argument in this direction rests on the assertion that quantum theory, as it stands, is incomplete: *if quantum theory were complete, then no constructive explanation would be possible because it fails to describe anything moving through ordinary three-dimensional space.* This line of reasoning had already been put forth by thinkers like de Broglie, Lorentz, and Einstein in opposition to Schrödinger's framework, as we saw. It is interesting to note that Heisenberg also found it untenable to accept a theory that posited no fundamental physical fields or space in three dimensions. He reportedly said of Schrödinger's theory: "Nonsense, [...] space is blue and birds fly through it" (Bloch 1976).

As previously noted, Schrödinger responded to such concerns by proposing a three-dimensional charge-density field ontology. Although this was a step toward incorporating a more realistic framework, it remained problematic since it still predicted macroscopic superpositions as well as the spreading of the wave packet. Nevertheless, Schrödinger continued to resist the inclusion of de Broglie's notion of particles, even though doing so would have at least addressed the first of these problems.

1927–35: Nonlocality Objection

While Heisenberg was increasingly drawn toward positivism due to the unphysical field objection, Einstein began searching for a spatiotemporal ontology, which he called *"hidden variables," to complete quantum theory*. In 1927, he proposed a deterministic particle theory, but later he arguably abandoned it because it turned out to be nonlocal. In systems composed of two independent parts, the theory failed to assign independent velocities to each (see Belousek 1996, Holland 2005). For Einstein, who was deeply committed to both constructive explanation and locality, this nonlocality rendered the theory unacceptable.

Unable to find other alternatives, Einstein spent the years between 1927 and 1935 developing several arguments based on the idea that *nonlocality was evidence of the incompleteness of quantum mechanics*. He aimed to convince others that this was the crucial area of research to focus on. One of the thought experiments he proposed in 1909 at a congress in Salzburg (Bacciagaluppi and Valentini 2009, p. 198) and later discussed at the 1927 Solvay Conference was as follows: *A particle travels toward a narrow slit* with a curved detection screen behind it. After passing through the slit, the particle's wave function spreads spherically due to diffraction and should hit every point on the screen. Yet experimentally, only a single localized detection event occurs. To Einstein, this was evidence that a nonlocal process must be selecting one point over others. Since such nonlocality was, in his view, unacceptable, he concluded that quantum theory must be incomplete (see Bacciagaluppi and Valentini 2009, p. 440). Schrödinger presented a remarkably similar example: suppose a radioactive source emits particles in all directions, one at a time. If the wave function provides a complete description and evolves according to the Schrödinger equation, then the particle must be detected simultaneously in all directions, which is not the case. Again, the only options are either incompleteness or nonlocality.

Bohr (1928, 1949) misinterpreted Einstein's argument, mistakenly believing it to be about the uncertainty principle. This confusion likely stemmed from Einstein's own framing, where he emphasized that the particle's position becomes known upon passing through the slit, while its momentum becomes fully uncertain. Possibly, the argument would have been clearer had Einstein omitted any mention of momentum, as in the reconstruction above. Nevertheless, Heisenberg understood Einstein's intended point. In his 1929 Chicago lectures, Heisenberg acknowledged that quantum theory entails nonlocality. He described a case where a wave packet is split by a half-silvered mirror, creating reflected and transmitted components that move to distant regions. If the reflected component is detected, then the transmitted one must vanish instantaneously, thereby violating locality (Heisenberg 1949, p. 39). Still, Heisenberg argued that no contradiction with relativity arises as long as the theory is understood instrumentally, as a theory of signals, since this form of nonlocality cannot be exploited for faster-than-light communication.

At another Solvay meeting in the 1930s, Einstein offered a revised thought experiment. Imagine *a photon inside a box*. After it is released and travels far away, the photon remains in a superposition of energy states. However, if someone measures the box's weight, the energy state of the photon collapses instantly, even though it is now distant from the box. Once again, locality is violated. Yet again, Bohr misunderstood the point, interpreting it through the lens of the time-energy uncertainty relation. Einstein had described the thought experiment in terms of the ability to measure either the photon's energy or the time with great accuracy, which led Bohr to deliver what many considered a devastating rebuttal using Einstein's own relativity theory against him. In reality, however, *they were simply talking past each other*. Einstein's core concern was nonlocality, not uncertainty (see Howard 2007, Nikolic 2012 for insightful discussions).

Einstein introduced another scenario, sometimes called the "*Einstein boxes*" experiment, in a 1935 letter to Schrödinger written prior to the cat paper: A box filled with a wave function is split in two by inserting a partition, which divides the wave function as well. These two halves are then separated over a great distance. If one half is opened, the state of the other is instantaneously influenced, which again implies a violation of locality (Einstein to Schrödinger, June 19, 1935, in Fine 1986, p. 35).

This line of argument culminated in the 1935 *Einstein-Podolsky-Rosen (EPR) paper*. EPR claimed that if quantum theory were complete, then measuring one particle from a pair of entangled particles (moving in opposite directions) would instantaneously determine the state of its partner, no matter how far apart they are. According to quantum mechanics, particles have no definite properties before measurement. Instead, these properties are created by the measurement process. However, by making a measurement on one particle, *we instantaneously create the property of the partner particle*, which could be at an arbitrary distance. If that were the case, then the theory would *blatantly contradict locality*, which was unthinkable. Therefore, EPR concluded, quantum theory must be incomplete. There must exist hidden variables that encode the particles' actual properties prior to measurement, and these properties are revealed, rather than created, by the measurement process. Unfortunately, this argument was also misunderstood by Bohr. Since the EPR discussion involved measurements of position and momentum, Bohr responded as if the argument were another critique based on the uncertainty principle. Consequently, Einstein was portrayed by the increasingly dominant Copenhagen school as an old-fashioned scientist unwilling to accept the quantum revolution. He was depicted as someone clinging to obsolete ways of thinking, rather than a forward-looking and open-minded visionary, ready to embrace the groundbreaking consequences of quantum physics.

1932: Impossibility Proofs

In addition to these misunderstandings, there were more legitimate reasons for abandoning the effort to complete quantum theory by introducing hidden variables. In fact, several theorems were developed with the explicit goal of proving the impossibility of such a program. One of the most notable was von Neumann's work in 1932, in which he claimed to have proven that *hidden variables could not exist*. He assumed, in line with the EPR argument, that certain properties exist prior to measurement and are revealed in experiments as the eigenvalues of quantum operators. Under what he considered reasonable assumptions, he demonstrated that these pre-existing properties would lead to mathematically inconsistent relationships. From this, he concluded that quantum theory could not be completed, and thus could not offer the kind of constructive understanding that had traditionally been the hallmark of physics.

Even though von Neumann's proof was later shown to be flawed—since some of his assumptions were not physically justified—other theorems were later proposed that did not rely on his problematic assumptions (Gleason 1957, Kochen and Specker 1967). However, *these theorems still assumed that genuine measurement could occur*. In reality, the interaction between the measuring device and the system is often sufficient to disturb the system to such a degree that the measurement outcome cannot be considered the value of the property prior to the experiment. In other words, these theorems did not prove that hidden variables were impossible; rather, they showed that *experiments often fail to reveal any intrinsic properties of the system at all*. This subtlety was fully appreciated only later, when Bell exposed the flaw in von Neumann's reasoning in 1964. Interestingly, Grete Hermann had already identified this issue in 1935, but her critique was completely ignored (Hermann 1935; see also Bacciagaluppi and Crull 2016). Einstein, too, dismissed von Neumann's result in 1938 as unphysical, although he never published his objections (Shimony 1993).

Nevertheless, *von Neumann's impossibility proof was widely accepted as a definitive refutation of the hidden variables approach* to quantum theory—largely due to his towering reputation, but presumably also because Einstein (or somebody else) did not respond to it in print for a long time. The sense of inevitability surrounding the Copenhagen interpretation reached its peak with this proof, which was seen as conclusively demonstrating that no alternative could provide a better explanatory framework than quantum theory. The hope for a constructive account of quantum phenomena, grounded in a deeper theory, was believed to be irretrievably lost.

1935: The Measurement Problem

The objections to the completeness of quantum theory raised by Schrödinger and Einstein focusing on unphysical ontology and nonlocality did not succeed in convincing anyone, so *they gradually adjusted their arguments to the target*. As a result, their objections became progressively weaker. The first type of argument, which was based on the need for a three-dimensional ontology to preserve a constructive explanatory framework, proved ineffective. This was largely due to the Copenhagen school's positivist stance, which led them to regard such constructive explanations as misguided because they relied on unobservable entities. This resistance effectively undermined any attempt to challenge the theory by insisting on a representational model grounded in physical space. Subsequently, arguments centered around nonlocality were developed to show that if quantum theory were complete, it would violate locality. However, these arguments also proved to be ineffective, as the Copenhagen school's instrumentalist interpretation of theories did not view nonlocality as a real problem. In their view, nonlocality could not be exploited to transmit information faster than light, and thus, it did not violate relativity.

Arguably, the final line of attack focused on what the instrumentalists themselves seemed to prioritize: empirical predictions. *If one could demonstrate that quantum theory failed to account for observable phenomena, then perhaps the Copenhagen group would take notice*. This, I believe, is the core motivation behind Schrödinger's famous cat thought experiment (1935a). If anti-realists were not troubled by microscopic superpositions, such as a decayed-and-not-decayed nucleus, because they are not directly observable, they would have to be concerned when a predicted macroscopic superposition—like a dead-and-alive cat—is not actually observed. In the weeks leading up to the publication of the cat paper, Einstein discussed with Schrödinger a remarkably similar scenario. In a letter, Einstein proposed an example involving a chemically unstable substance, such as a charge of gunpowder, which might spontaneously combust through its internal forces. If it combusts, it triggers an explosion; if it does not, nothing happens. The substance would start in a superposition of "combust" and "not combust," which, through amplification, leads to a superposition of "explosion" and "no explosion." Yet, such superpositions are never observed. Einstein wrote:

> The system is a substance in chemically unstable equilibrium, perhaps a charge of gunpowder that, by means of intrinsic forces, can spontaneously combust, and where the average life span of the whole setup is a year. In principle, this can quite easily be represented quantum-mechanically. In the beginning, the ψ-function characterizes a reasonably well-defined macroscopic state. But, according to your equation, after the course of a year, this is no longer the case at all. Rather, the ψ-function then describes a sort of blend of not-yet and already-exploded systems.
> (Fine 1986)

This directly anticipates the scenario Schrödinger would articulate more formally a few months later.

Therefore, this line of reasoning aimed to show that *if quantum theory were complete, it would not be empirically adequate.* The problem it raises is now known as the measurement problem: if the wave function is taken to completely describe physical systems and evolves according to Schrödinger equation, then all possible outcomes of an experiment are predicted to occur simultaneously. Yet, in reality, only one outcome is ever observed.

Still, *this argument did not succeed either*. By 1932, von Neumann had formulated a response: his *collapse postulate*. That is, the wave function evolves deterministically via the Schrödinger equation except when a measurement occurs; at that moment, the wave function collapses spontaneously and randomly into one of the terms of the superposition. This collapse is nonlocal, but that was accepted, especially since, as Heisenberg emphasized, it cannot be used to send signals or transmit information. As a result, the Copenhagen interpretation once again prevailed, and the measurement problem and related conceptual issues were largely ignored until the 1950s, when concerns about the lack of a unified dynamics began to reemerge.

Meanwhile, the onset of the Second World War scattered the scientific community and distracted it from foundational concerns. The war interrupted collaborations, conferences, and philosophical debates, while shifting the focus decisively toward technological applications rather than conceptual clarity. In this climate, most physicists uncritically adopted the mainstream Copenhagen view, raising few challenges or objections.

1950s: Bohm's Rediscovery of de Broglie's Theory

Gradually, by the 1950s, things began to return to normal. After writing a textbook on quantum theory, Bohm (1952) started to reflect more deeply on the subject, ultimately revisiting (unknowingly) de Broglie's earlier work and expanding upon it. While at Princeton, he proposed a theory in which particles evolve according to an equation derived from a Schrödinger-evolving wave function. *Bohm developed this theory as a constructive understanding of quantum mechanics, rather than as a solution to the measurement problem.* As already mentioned, Bohm formulated Newton's equation of motion and demonstrated that quantum predictions could be reproduced by introducing a quantum potential, which is expressed in terms of a Schrödinger-evolving wave function. In this model, the wave function never collapses, and there are no macroscopic superpositions. Instead, matter is composed of particles, and the only aspect of the wave function that influences their motion is the one corresponding to the particles themselves, effectively allowing the other terms to be disregarded for practical purposes.

Initially, Einstein encouraged Bohm to continue pursuing this theory, but Bohm was largely ostracized by the rest of the scientific community for challenging the Copenhagen dogma. Moreover, Bohm's political sympathies toward communism led to his exile in Brazil, where he was left without a passport and unable to travel to defend his theory against attacks from the Copenhagen school. After several years, mostly spent in exile, Bohm essentially stopped advocating for his theory.

While Einstein was initially supportive of Bohm, he grew *critical of the theory's explicit nonlocality*. He also disliked the fact that, like Schrödinger's wave mechanics, the pilot-wave theory's *wave function is not three-dimensional*—a feature that Einstein believed made no physical sense. Thus, Einstein became convinced that quantum theory needed to be completely restructured, rather than merely supplemented. He wanted to preserve both locality (due to its intuitiveness and the constraint of relativity) and the three-dimensionality of the ontology (to maintain a constructive explanation). However, he could not find any proposal for a constructive theory that was also local. This is why Einstein was drawn to field theory: he viewed three-dimensional fields as mediators of interaction, which could propagate at finite velocities, thus maintaining locality, and possibly grounding a constructive explanation.

1960s: Contextuality of the de Broglie-Bohm Theory

Most people dismissed the pilot-wave theory, aside from Einstein and Schrödinger. In light of von Neumann's alleged proof that completing quantum theory was impossible, they believed there had to be something fundamentally wrong with the theory, even if they could not clearly pinpoint what it was.

Even those who were open to the idea that Bohm's theory could bypass the impossibility proofs still believed that it came at a significant cost. It seemed that the theory would require the *postulation of contextual properties*, which they considered a major flaw. Traditionally, when we think of a property of a system, we regard it as a stable, intrinsic characteristic, something that is timeless. For example, if we say the spin along the x-axis is "up," this is considered a fixed property of the system. When we set up an experiment to measure this property, we expect the experiment to yield the same result each time, under identical conditions. If we then repeat the experiment after measuring another property, say the spin along a different direction, the value of the spin along the x-axis should remain unchanged. However, critics of the pilot-wave theory argued that, in this model, properties are not stable in this way. Instead, they depend on the context in which the measurement occurs. In other words, if we measure the spin along the x-axis first and then measure it along another direction, the value of the spin along the x-axis could change depending on the order of the measurements. This contextuality in the de Broglie-Bohm theory was seen as a significant disadvantage, suggesting that in attempting to frame a microscopic picture of the system, we are forced to deal with strange and counterintuitive properties that don't align with our classical understanding.

However, as Bohm himself clarified in his 1952 paper, while his theory was often misunderstood, it does not fall prey to the impossibility proof. This is not because it postulates contextual properties, but because it operates in line with Heisenberg's uncertainty principle when properly understood. Bohm argued that when an experiment is performed on a system, it inherently alters the system. This interaction between the experiment and the system means that subsequent measurements should not be expected to yield the same results as the initial measurement, even if the same property is being measured. In other words, Bohm's theory acknowledges that the very act of measurement disturbs the system in a way that makes the results of repeated measurements inconsistent, which aligns with the uncertainty principle. This understanding offers a coherent explanation for the perceived contextuality in the de Broglie-Bohm theory, suggesting that it is not a flaw but rather a natural consequence of how measurements affect physical systems. Nevertheless, this response went unheard, for the reasons we anticipated, and the theory continued to be disregarded for decades.

1960s–80s: Bell's Inequality

After observing Bohm seemingly "doing the impossible" by circumventing the impossibility proofs, *Bell demonstrated that the impossibility proofs mistakenly assumed one could measure properties without altering the system* (his paper, written in 1964 and published in 1966). *The pilot-wave theory*, which assumes that experimental apparatuses may alter the system, *bypasses these impossibility proofs and completes quantum theory, but at the cost of being nonlocal.* Bell then began to consider *whether it was possible to complete quantum theory while preserving locality*, as Einstein had hoped. For this reason, in 1964, he revisited the EPR argument: the perfect correlations observed in entangled systems can only be locally explained by assuming the existence of properties whose values are not created by the measurement but that exist even before the measurement. By assuming that measurement reveals rather than creates these values, Bell derived an inequality that holds for this theory but not for quantum mechanics, providing a crucial test. This test was later conducted by Aspect and collaborators in 1982, confirming quantum mechanical predictions.

From this result, Bell wanted to conclude that any theory that matches quantum predictions must be nonlocal. However, *most people mistakenly thought he had proven that hidden variables*, as envisioned by Einstein (whether local or not), *are impossible.* While this is true in the sense that experiments do not reveal pre-existing properties, it was not Bell's aim to prove this point. In fact, he already knew that von Neumann's theory was unsound: since there are no genuine measurements, pre-existing properties are not revealed by measurements and thus cannot explain nonlocal correlations. Rather, he demonstrated that nonlocality is necessary to reproduce quantum predictions. However, the confusion was compounded by the fact that Bell started from the EPR argument, which assumed what Bell knew was false—namely, that experiments could reveal pre-existing properties of a system. To clarify this misunderstanding, Bell later (1975) reformulated his inequality without relying on the EPR argument, assuming only the locality condition. This reformulation shows that any theory matching quantum mechanical predictions must be nonlocal. However, controversies remain about what Bell's work ultimately proved.

1950s–Present: The Solutions of the Measurement Problem

Research into the meaning of quantum theory barely continued in an organized manner after Bohm was ostracized. While he had sought to offer a constructive understanding of quantum phenomena, *Bohm's theory was effectively shut down by the impossibility proofs*. Starting in the late 1950s, the attention of those still interested in understanding the theory slowly shifted from the problem of completing quantum theory to provide a constructive description from a microscopic perspective, to *addressing the measurement problem*, which focuses on macroscopic superpositions.

Some researchers began to question the orthodoxy, which now included von Neumann's collapse postulate, not because of its non-three-dimensional ontology or its nonlocality, but from within the established dogma itself. For example, Everett, a student at Princeton, argued that *von Neumann's collapse was unsatisfactory*. Unlike Schrödinger, Einstein, de Broglie, and Bohm, Everett was not motivated by realism. Instead, he was concerned with the imprecision in von Neumann's formulation and the inconsistencies he argued followed from it. In his 1957 dissertation, he proposed an original way to make sense of macroscopic superpositions. He suggested that the various terms in the superposition represent relative states of affairs for an observer. One could think of the wave function as representing the point of view of various observers, each corresponding to a different term in the superposition. His advisor, Wheeler, naively enthusiastically discussed this idea with Bohr, who, however, did not react positively. This ultimately contributed to Everett's career being effectively destroyed, and after a very short while, he left academia to work in military research. Feynman, who heard about the theory from Wheeler, found this idea preposterous. This approach was forgotten for years, until DeWitt, a cosmologist and friend of

Wheeler, heard about it, and in 1970 published a paper on quantum theory and reality. He is the one who changed the name of the theory into "the many-worlds interpretation of quantum mechanics." The idea is that the superposition, instead of representing observers, can well be thought of as representing a world. This article was followed by the DeWitt and Graham book on this theory in 1973, in which they addressed the problem of cosmology, specifically asking where the observer is if the system is the entire universe. In 1985, Deutsch, who had attended an Everett lecture in 1977, published a paper on *quantum computing*, arguing that quantum computers are faster than classical ones because there is a world for each branch of the wave function where the calculations happen. The many-worlds theory, unlike Bohm's approach, did not add anything new to the standard theory and did not modify the Schrödinger equation (unlike collapse theories, which would later be discussed). Instead, it seemed to many to be the best compromise among the solutions to the measurement problem, assuming one disregards the potentially inflated metaphysical implications, which, for most physicists, arguably appeared a "romantic" price to pay for mathematical simplicity. As a result, the many-worlds theory began to gain traction, particularly among *cosmologists* who had already realized that the Copenhagen interpretation was inadequate, and among those working on quantum computation after Deutsch's results on quantum computers. One challenge for the many-worlds theory is to show that the different worlds do not interact with one another, as this would violate the principle of observation. *Decoherence theory*, which began to be developed by Zeh in the 1970s, provided a solution. Decoherence suggested that physical systems interacting with their environments lose coherence, and thus their ability to interfere with themselves.

The measurement problem was also explored by Wigner (1961), who believed that the von Neumann collapse was real and triggered by *consciousness*. This idea was immediately attacked by Rosenfeld, a staunch defender of Bohr's views (see Freire 2015).

In the 1970s, the first experiments testing Bell's inequality were conducted, with the violation of Bell's inequality confirmed in 1982. Furthermore, the first conferences on quantum foundations, all focusing on the measurement problem, were held in 1970 in Varenna and in 1973 in Erice.

During the same period, a few researchers criticized von Neumann's collapse postulate because it introduced two fundamental evolutionary laws. They worked on *incorporating the collapse process into the dynamical equation of the wave function,* aiming to provide a unified dynamics for both the macroscopic and microscopic realms. Their work was more aligned with Everett's approach than with that of Einstein, Schrödinger, de Broglie, or Bohm, in that they did not focus on the ontology or the nonlocality of the collapse. Instead, they sought *to make the Copenhagen recipe more precise*. Unlike Everettians, however, they did not think that introducing a multiplicity of descriptions to accommodate superpositions was a nonstarter. Rather, they aimed at suppressing superpositions when they are observable, and keeping them when they are not. This had to happen as a matter of law rather than as a matter of observation. This effort culminated in the dynamical collapse models proposed independently by Pearle and by Ghirardi, Rimini, and Weber in the early 1980s. In this theory, the wave observed. Furthermore, the wave function of a system with multiple "particles" does not vibrate in three-dimensional space, but rather exists in *configuration space*—a higher-dimensional space. This raised a problem: if the ontology of the theory does not reside in three-dimensional space, then constructive explanation fails. At the 1927 Solvay conference, de Broglie justified his introduction of particles by stating: "it seems a little paradoxical to construct a configuration space with the coordinates of points that do not exist" (Bacciagaluppi and Valentini 2009, p. 346). Lorentz, writing to Schrödinger, said:

> If I had to choose now between your wave mechanics and the matrix mechanics, I would give the preference to the former, because of its greater intuitive clarity, so long as one only has to deal with the three coordinates *x, y, z*. If, however, there are more degrees of freedom, then I cannot interpret the waves and vibrations physically, and I must therefore decide in favor of matrix mechanics.

Similarly, Einstein wrote to Lorentz on May 1st, 1926: "Schrödinger's conception of the quantum rules makes a great impression on me; it seems to me to be a bit of reality, however unclear the sense of waves in n-dimensional q-space remains" (Lorentz in Przibram 1967). (Here, q-space refers to configuration space.) Einstein also expressed his concerns in a letter to Ehrenfest on June 18th, 1926, stating: "Schrödinger's works are wonderful—but even so one nevertheless hardly comes closer to a real understanding. The field in a many-dimensional coordinate space does not smell like something real." (These quotes from Einstein are taken from Howard 1990.)

Schrödinger was aware of the spreading wave packet problem but was unable to find a solution. He also recognized that interpreting a field in high-dimensional space did not align with the constructive understanding of the world. In 1926, Schrödinger noted: "The direct interpretation of this wave function of six variables in three-dimensional space meets, at any rate initially, with difficulties of an abstract nature." He also wrote: "Of course this use of the q-space is to be seen only as a mathematical tool, as it is often applied also in the old mechanics; ultimately [...] the process to be described is one in space and time" (Bacciagaluppi and Valentini 2009, p. 447). To address these issues, he proposed a three-dimensional field ontology as the fundamental ontology of matter, defined by the square modulus of the wave function integrated over a three-dimensional region, which represented the *charge density* of the "particle." He chose charge instead of mass to connect his theory with electromagnetism (1926). However, like all waves, Schrödinger's *charge density waves superimpose*, and these superpositions *propagate to the macroscopic domain* where, however, they are not observed. This led to the Schrödinger cat paradox.

What should one conclude when a theory predicts something that is *not observed*? One should at least conclude that the theory is not empirically adequate, and perhaps go so far as to claim that the theory has been falsified. In other words, if wave ontology leads to superpositions that are unobserved, and there is no independent reason to believe in it, then it seems reasonable to think that *the starting metaphysical hypothesis, which directly led to these predictions, is incorrect*. As is known, it is always possible to protect function evolves according to a stochastic, nonlinear equation, causing it to collapse at random times and locations, with the frequency depending on the system's size, so that macroscopic objects collapse almost instantaneously.

Following this proposal, Bell began to advocate for the idea that *there are three ways to make quantum theory compatible with scientific realism by solving the measurement problem*. As Bell put it in 1987, "either the wave function, as given by the Schrödinger equation, is not everything, or it is not right." Or, as more emphatically stated by Maudlin (1995):

> The following three claims are mutually inconsistent. 1.A The wave function of a system is complete, i.e. the wave function specifies (directly or indirectly) all of the physical properties of a system. 1.B The wave function always evolves in accord with a linear dynamical equation (e.g., the Schrödinger equation). 1.C Measurements of, e.g., the spin of an electron always (or at least usually) have determinate outcomes, i.e., at the end of the measurement the measuring device is either in a state which indicates spin up (and not down) or spin down (and not up).

Stating the problem like this officially legitimizes the many-worlds theory (which denies 1.C) and the dynamical collapse models (which deny 1.B) as acceptable proposals for realist quantum theories. Since most researchers who wondered about the meaning of quantum theory ended up addressing the measurement problem, they consider the de Broglie-Bohm theory, the Everett many-worlds theory, and the theory of dynamical collapse as *equally plausible* ways to make a realist sense of quantum theory, each with its own strengths and weaknesses. For instance, the pilot-wave theory is often criticized for its nonlocality, the many-worlds theory is questioned because of its handling of probabilities, and the spontaneous collapse theory is seen as *ad hoc*.

Moreover, since all these theories incorporate the wave function, it became natural to interpret it as the ontology of the theory, embracing what was initially thought impossible: a *high-dimensional ontology*. One might argue that this is the most straightforward way to interpret these theories: they all have an equation for the wave function, so the wave function must be the ontology (Albert 1996). Alternatively, one could argue that this is the only way to restore locality to the theory, even if in configuration space: Bell demonstrated that reality is nonlocal, but if we adopt an ontology in configuration space, we can still maintain that type of locality (Ney 2021). This view has gained popularity and is now known as wave function realism, which we discussed in Chapter 3. This growing field has further undermined the pilot-wave theory, as it does not fit this framework. If one sees the theory as having a wave function ontology existing in a high-dimensional space, one must either assume that there is also a three-dimensional space where particles reside or that there is only one particle living in the same space as the wave function. In both cases, the theory's depiction of the world becomes highly revisionary, leading many to question the necessity of postulating particles at all (Brown and Wallace 2005). The point, however, is that particles (or a microscopic spatiotemporal ontology) are essential if one desires a constructive explanation of phenomena, something easy to overlook when focusing on solving the measurement problem.

The Measurement Problem Is a Wild Goose Chase

In the previous section, I argued that *the measurement problem is a concern for anti-realists who prioritize empirical adequacy*. While a satisfactory theory must address this problem, resolving it does not necessarily ensure that the theory can offer a constructive explanation of the phenomena. But *which argument should realists have focused on*, if not the measurement problem? In this section, we explore how the various arguments for incompleteness discussed in the last section differ from one another. The unphysical field arguments claim that quantum mechanics is incomplete because there is nothing moving in three-dimensional space. Instead, nonlocality arguments argue that quantum mechanics is incomplete otherwise it would be nonlocal. Finally, macroscopic superpositions arguments state that quantum mechanics is incomplete because it predicts unobserved macroscopic superpositions.

As discussed, the first two arguments were systematically ignored (such as the unphysical field problem) or misunderstood (such as the nonlocality issues). Ironically, the only argument that garnered attention later by realists, namely the measurement problem, was initially proposed as a last resort to convince the positivists that something was missing in their theory.

The unphysical field problem is the most stringent of these arguments. It was proposed assuming that explanation needs to be constructive, and it is shown that in order to have one, the theory needs to be completed with a three-dimensional ontology. This argument starts from the assumption that the wave function cannot be interpreted as a physical field vibrating in space because it is not in spacetime. The nonlocality arguments focus on the fact that if quantum theory is not completed in one way or another, locality is lost. While the first argument tells us what to do (add a spatiotemporal ontology), *the locality argument does not explain how to complete the theory*. Rather, it merely states that it must be done. Presumably, this shift in focus occurred because members of the Copenhagen school were less concerned with the microscopic description and more focused on the results. However, as already mentioned, Heisenberg did not mind the nonlocal nature of quantum theory. Therefore, the last attempt came in 1935, when the incompleteness argument was framed in macroscopic terms, which were the only ones the positivists would consider: if quantum mechanics were complete, it would predict unobserved macroscopic superpositions and thus it would be false. Thus, we need to eliminate these superpositions, regardless of whether the theory is local or not, or whether it is completed with a three-dimensional ontology or not. Therefore, among all these arguments, *the measurement problem is the weakest*. It allows for *ad hoc*, nonlocal, and imprecise solutions, such as the von Neumann collapse rule.

Therefore, realists who care about constructive explanation should not be content in solving this problem: not all solutions of the measurement problem are acceptable, only constructive ones will work. That is, *realists should focus on ensuring a three-*

dimensional ontology from which they can construct an explanation, and then the task becomes finding the correct ontology. The natural choice would be to decide between particles and fields, as had been done before. Take a field ontology, like Schrödinger's charge density proposal. Even assuming we could control the field packet's spreading through decoherence, the theory still shows unobserved macroscopic (three-dimensional) superpositions. It's as if we postulate that matter is a wave, but we do not observe interference. What should we conclude? One possibility is that we are mistaken, and matter is fundamentally not a field. The best way to resolve this problem is by adding particles, which is precisely what the de Broglie-Bohm pilot-wave theory does. The alternative is to stubbornly insist that matter is a field and adopt a many-worlds interpretation to explain why we do not observe macroscopic superpositions, even if they exist. In this interpretation, each term of the superpositions exists in different worlds that do not interact. However, this second option appears completely *ad hoc* and unnecessary. It seems like using the Quine-Duhem thesis to maintain our favored ontology. The first option, particles, offers a simpler, more sensible, less problematic, and less revisionary solution. Before the advent of quantum weirdness, *no one would have dared propose anything like this.* Any such proposal would have been seen as worse than denying the reliability of telescopes, almost like suggesting the existence of undetectable ghosts to explain particle behavior when forces fail to do so. In contrast to the second option, the first option preserves the constructive schema. Schrödinger and Einstein's goal was to find a constructive theory to ground quantum mechanics, which had initially been a principle theory. They sought a corresponding "statistical mechanics" for "thermodynamics." This is what the first option provides. On the other hand, many-worlds provides a more complex principle theory, where the principles apply to unobserved and unobservable worlds. What is the advantage of that? Proponents of many-worlds argue that it best respects the practice of physicists: they do not use collapse, nonlinear stochastic dynamics, or additional variables; they simply use Schrödinger evolution. But what accounts for this practice? Why is this the approach physicists adopt? As discussed in Chapter 4, the pilot-wave theory offers an explanation beyond "just because:" one does not use von Neumann's collapse rule because the system is effectively guided by the conditional wave function, which either evolves linearly or collapses when necessary; one uses operators because they mathematically summarize the results, even in circumstances where experiments do not reveal a property of the system; one does not need to modify the Schrödinger equation to avoid superpositions because the system consists of particles that do not superimpose, and so on.

Are there other alternatives? I do not think so, especially if we care about the constructive schema. The other proposed solution to the measurement problem is to modify the Schrödinger equation (stochastically and nonlinearly). While this provides a more elegant and unified principle theory than standard quantum mechanics, it does not offer a constructive theory. If one were to seek a constructive theory in this context, one would need to add a three-dimensional ontology to the spontaneous localization dynamics. But why complicate the wave function's evolution if we can achieve the same results as Schrödinger's evolution using the pilot-wave theory?

Needless to say, if someone believes that wave function collapse is not real (as in the pilot-wave theory or many-worlds approach), or that it is dynamic (as in spontaneous localization theories), there is no reason to assume that physics requires consciousness to provide a complete description of reality.

Thus, I think that by *focusing on the measurement problem*, namely the issue of suppressing unobserved macroscopic superpositions, *one gives too much credit to ideas that would have never been considered had history unfolded differently.* Realists should have focused their efforts on preserving the constructive schema as far as it was possible. This is why, in the 1920s, some individuals (Einstein, Lorentz, de Broglie, etc.) argued that quantum theory needed a spatiotemporal ontology. In this framework, the concept of a wave function ontology would never have emerged, along with the many-worlds theory and the spontaneous collapse model, with the pilot-wave theory being the only surviving proposal.

Preserving High-Dimensional Locality

The locality argument discussed above asks us to complete the theory in a local way. However, Bell showed that we cannot have that: nature is nonlocal, and the pilot-wave theory embraces this nonlocality in the most straightforward way. So, we can keep the constructive explanatory schema in terms of microscopic entities moving around in three-dimensional space and composing macroscopic bodies. However, as an empirical fact, there are nonlocal correlations between these entities.

What if we wanted to preserve locality? A proposal is given by *wave function realism*, as seen in Chapter 3, so I will not reiterate it here. In a nutshell, by postulating that the true ontology is the wave function in high-dimensional space, it is argued that one *can preserve locality in the fundamental space*. As we discussed, there are various reasons to require locality from a physical theory: it is compatible with relativity, it makes sense of the notion of interaction as traveling, it is compatible with the idea that causes precede their effect, and it allows us to think of objects as isolated, thus making sense of constructive explanation. Wave function realism, however, preserves locality in the fundamental space, which is a high-dimensional space, not spacetime. So, it is unclear why one would want to preserve locality in that space, when all our arguments were for locality in spacetime.

Preserving Locality with Superdeterminism or Retrocausality

If what we really care about is *locality in three-dimensional space*, we need to address the empirical violation of Bell's inequality. One way to do this is to *question* a commonly accepted assumption in Bell's derivation: *statistical independence*. This principle holds that the outcome of a measurement is independent of the experimenter's choice of which measurement to perform. Statistically speaking, two events are independent if the occurrence of one does not influence the probability of the other. In our context, this means the outcome of the measurement should not be influenced by the experiment that the observer has chosen to conduct. Let's take an example. Imagine you have a drawer containing two coins, one copper and one silver. You can choose to test either the material (copper vs. silver) or the thickness (thick vs. thin) of the coin you pick. Suppose you decide to test the material. If the probability of finding a silver coin is ½, repeating the same type of measurement over many trials should result in roughly half silver and half copper outcomes. Crucially, the choice to measure thickness instead should not influence the distribution of silver vs. copper outcomes, since the material and thickness are intrinsic features of the coin, independently of your choice of measurement.

What has been called "*superdeterministic*" theories, by contrast, assert that statistical independence does not hold. According to them, your choice of what to measure and the outcome you observe are such that this independence is lost. This opens the possibility of reconciling locality with the violation of Bell's inequality. But *is it worth giving up statistical independence just to preserve locality*? To explore this, we need to understand the implications of rejecting it. Returning to the coin example, in a superdeterministic framework, if you decide to measure material and you get "silver," this result doesn't necessarily reflect the intrinsic property of the coin. Instead, the result you get is tied to your decision in a deeply constrained, predetermined way. Over many trials, even though you think you are randomly selecting coins and measurements, you would always end up picking the silver coin when you choose to measure material, and never the copper one. Your outcome and choice together "super-determine" the object you are actually interacting with. Thus, the probability of getting "silver," given that you chose to measure material and got "silver," becomes 100 percent. You never learn anything about the broader distribution of copper or silver coins, because you are always steered toward coins whose properties match your measurement outcomes. Aside from the fact that we lack a fully developed superdeterministic theory (though see Ciepielewski et al. 2021), a major concern is that *rejecting statistical independence seems to undermine the very foundation of empirical science*. Experiments aim to uncover information about the systems under investigation, but superdeterministic theories deny this possibility. Measurement no longer reveals anything fundamental. Many scholars (e.g., Maudlin 2019b, Baas LeBihan 2021, Chen 2021, Allori 2024a) argue that statistical independence is indispensable for the scientific method. Furthermore, there *is no satisfying account of why* your decision about what to measure would dictate what system or object you actually interact with. This explanation appears far more obscure than accepting nonlocal influences. Without an independent, non-*ad hoc* justification, superdeterminism seems like a contrived solution aimed solely at salvaging locality at all costs.

In contrast, *retrocausal theories* strike me as more promising. These theories, such as the transactional interpretation (Cramer 1986, Kastner 2022), the two-state vector formalism (Aharonov, Bergmann, and Lebowitz 1964), the recent approach of Price and Wharton (2023), and the relational block-world view, also violate statistical independence, but they do so by introducing backward-in-time causation. Broadly speaking, in these models, A causes B if manipulating A allows one to influence B. Retrocausal frameworks suggest that changing the measurement setting in the present can influence aspects of the system's past. I find retrocausal theories most intuitively understood through a block-world perspective, in which spacetime is a four-dimensional structure traced by the worldlines of particles. Theories like classical mechanics and pilot-wave theory also fit into this block-world view, where the worldlines are determined by past events and evolve forward. Retrocausal models extend this by allowing future events to influence these worldlines as well, thus creating a symmetrical relationship between past and future. Therefore, within this framework, they are rightly called "all-at-once" theories (Adlam 2022). Theoretically, this time symmetry is elegant. However, since we don't observe this symmetry directly in experience, any such theory must explain why this is the case.

The next relevant question for my perspective is *whether retrocausality can be integrated into the constructive explanatory framework* developed in this book. Presumably, it can—provided such models are treated as dynamical theories rather than principle theories. From a block-world perspective, adopting principle theories might seem natural, as they can be viewed as imposing constraints on spacetime trajectories. Nonetheless, this is not the only way to understand them. While any view can be adjusted to suit one's theoretical preferences, the real question is whether such adjustments are genuinely plausible. In any case, what appear to be constructive retrocausal pilot-wave models have been proposed and deserve further exploration (see, for example, Sutherland 2008). From my perspective, then, the relevant comparison is between nonlocal theories such as traditional pilot-wave mechanics and local but retrocausal alternatives. I cannot predict whether retrocausal models will ultimately succeed, but I do think they offer a viable and theoretically appealing path, certainly more so than some of the alternatives, and they deserve far more serious attention than they have so far received.

Preserving Constructive Explanation over Locality

If we care about theories being in spacetime, and if we do not want superdeterminism or retrocausation, then the natural remaining option is the pilot-wave theory. This is because, as I have argued in Chapter 4, the stochastic alternatives are too costly: they are *ad hoc*, they violate the quantum predictions, they threaten the notion of interaction, and they are more nonlocal than needed. As we have seen, pilot-wave theory can be made Lorentz invariant, assuming we add a preferred foliation. As noted, while it is true that one must add it to the relativistic spatiotemporal structure, it is neither *ad hoc* nor unnecessary. And although the preferred foliation is unobservable, this can be accounted for using quantum equilibrium: we do not have access to the particle's configuration. If we did, namely if we were outside of equilibrium, we could observe the preferred frame (Valentini 2007). Moreover, given quantum nonlocality is a fact, the preferred foliation allows us to account for it straightforwardly. So, it is explanatorily necessary, rather than redundant.

Naturally, there are many open questions. Quantum nonlocal interactions are instantaneous. What implications does this have for the notion of interaction? As we saw, the idea of locality captures the idea that interaction travels: an object takes some time to "notify its presence" to another object. Since, instead, we have nonlocality, we are back to Newton's problem where lifting my arm instantly influences the motion of Saturn. As we saw, in the classical case we can alleviate the problem because we can use fields to mediate the interaction, which then becomes not instantaneous, and the strength of the interaction decreases with distance, so that we can ignore the effects of objects that are far away. In the quantum case, however, as anticipated, we do not have either: no spatiotemporal mediator, and no decrease of strength with distance (even if decoherence explains why we effectively never see nonlocality, see below). *Could (or should) we still think that interaction travels* under these constraints? If it does, it travels at a speed greater than the velocity of light, which in turn means that light is not the fastest speed. However, this was a consequence of Lorentz invariance: if signals could travel faster than light, then we could have signals arriving before they are transmitted, and I do not think the situation changes much with the introduction of a preferred foliation. If instead we give up the idea that interaction is something that travels, then what are the consequences, and how else should we understand interaction? As a side remark, let me notice that if we do abandon the notion of traveling interaction, we also lose the need for fields as mediators, which would make the overall ontology of the theory simpler. Be that as it may, *how should we think of entangled particles*? One possibility could be to think of them not as two particles interacting nonlocally, but as nonseparable entities. While the notion of locality has to do with interaction, the notion of separability has to do with the way matter is. That is, if the properties of the whole are completely determined by the properties of its parts, then we talk about separability. But in this context, matter is made of particles, so in virtue of what is a pair of entangled particles nonseparable? Another possible reaction is to simply maintain that the law governing the behavior of matter is nonlocal, and such a law does not "propagate," but rather it is what primitively connects a pair of entangled particles. Presumably, one can think of many more ways to make sense of nonlocality, and I do not know which attitude will prove to be most fruitful. But, since the word is nonlocal, not only those who endorse the pilot-wave theory but everyone else who is not a superdeterminist or a defender of retrocausation (who have to solve different problems) needs to start asking these types of questions and start finding answers.

To conclude this section, let me get to the question of *why nonlocality is hidden from us*. Einstein (like everyone before him, and many others even after Bell and Aspect's results) thought that the world was local; otherwise, it would be impossible to do physics: it seems necessary to talk about cause and effect and to treat systems as isolated. Now we know the world is nonlocal. So, how is it possible that we can do physics as if the world were local? The reason why we can effectively treat systems as isolated, and we can describe most of the interactions locally, is the same as the reason why it is so difficult to see entangled states in the lab: decoherence. Only if one really looks for it, with great difficulty, can one see nonlocal effects, as is clearly shown by the ingenuity of the experiments about the violation of Bell's inequality. In all other cases, for all practical purposes, we can ignore nonlocal effects because the interaction with the environment destroys the coherence between the subsystems, effectively making nonlocality unobservable in most circumstances.

Conclusions

In this chapter, I have told a fictional story: how quantum theory would have developed if rational arguments, rather than misunderstandings and power plays, governed theory selection. I called this world, "*the ideal quantum framework.*" Had we lived in such a world, we would have seen how to develop a constructive quantum theory which preserves symmetries, and we also would have identified what the true spirit of quantum theory is: nonlocality. This would have led us almost immediately to the pilot-wave theory, which is explicitly nonlocal. We could have seen straightforwardly how to think of relativity and quantum mechanics together. They are both incomplete: while quantum theory lacks a spatiotemporal ontology to ground constructive explanation, relativity lacks a preferred spatiotemporal slicing to account for quantum nonlocality. Relativistic pilot-wave theories supply both things in the simplest Lorentz invariant way. We would not have to deal with most of the things we read about in quantum foundations: quantum revolutions, many worlds, consciousness creating reality, and so on. Rather, we would have been left with the true problem: to understand what a nonlocal interaction amounts to.

Instead, *history has been different*, and we lost so much time, energy, and resources into research programs that, I argue, no one would have taken seriously in the ideal quantum framework. Hopefully, this little book will help change the course of the current world toward a deeper understanding of the quantum domain which does not require unnecessary revolutions.

Further Readings and Discussion

Relevant books on the history of quantum mechanics include Cushing (1996), Beller (2001), Bacciagaluppi and Valentini (2009), and Becker (2018). Notable is also the collection edited by Bacciagaluppi and Crull (2016) dedicated to Grete Hermann, one of the few women involved in quantum foundations. Invaluable collections of historical articles are van der Waerden (1968) as well as Wheeler and Zurek (1983).

On superdeterministic theories, see especially t'Hooft (2016), Hossenfelder and Palmer (2020), and Ciepielewski et al. (2021). For criticism, see Shimony et al. (1976), Maudlin 2019b, Baas and LeBihan (2021), Chen (2021), and Allori (2024a). A comprehensive review on retrocausality is given by Friederich and Evans (2022). See Adlam (2022) for a clear general discussion of retrocausal theories.

Epilogue

In this book, I have outlined what I take to be the most promising approach to quantum ontology. The central idea is that the purpose of physical theories is to provide *constructive explanations* of macroscopic phenomena, *grounded in spacetime* and consistent with *symmetry principles*. This is accomplished by focusing on the architectural structure of physical theories: at the base lies matter, fundamentally situated in spacetime, while the rest of the theoretical structure comprises elements necessary to build a constructive account of physical processes. The entities comprising the nomological ontology determine how matter behaves. This, then, is what I have called the *spatiotemporal foundational ontology* approach.

In classical physics, electromagnetic fields are best understood as nomological elements rather than as parts of the material ontology—especially if we aim to render the theory as perspective-free as possible, by accommodating the broadest possible range of symmetries. In the quantum context, the wave function plays a similar role. It, too, is best located within the nomological ontology, not only because it fails to be spatiotemporal, but also because its functional role mirrors that of classical fields: it governs motion. It is not a material field but a component of the law. Thus, within this framework, the wave function does not represent matter itself; it specifies how matter evolves. Hence the name I give this view: *wave-functionalism*.

Alongside this new understanding of the wave function, I propose a different conception of matter. I explore the idea that matter might be *bare*—lacking any fundamental properties beyond those necessary to individuate particles. That is, particles do not fundamentally possess mass, charge, or spin. Rather, *all particles are fundamentally identical*, and what we think of as their properties emerge from the laws they follow. In this approach, properties are not intrinsic to matter but are reinterpreted as aspects of the governing laws. This shift in perspective demystifies properties, emphasizes the importance of laws as structural relations among the fundamental material entities, and strengthens continuity with classical mechanics. Instead of assuming particles somehow acquire strange, context-sensitive properties in the quantum regime, we retain a stable particle ontology while recognizing that it is the laws that change.

Why do I emphasize *continuity in ontology*? Because it is essential to a realist interpretation of physical theories. Continuity helps resist the pessimistic meta-induction—the view that because many past successful theories turned out to be false, our current theories are likely to be false as well. But if the elements responsible for a theory's success endure through theory change, then we are justified in being realist about those elements—the *working posits*. In classical theories like Newtonian mechanics and Maxwellian electrodynamics, the successful elements include a particle ontology and deterministic laws. In the quantum domain, we find the same: a particle ontology and deterministic laws (as in the pilot-wave theory). The key difference lies in the form of the laws. In classical mechanics, motion is governed by forces mediated by fields in spacetime. In the quantum domain, motion is governed by a wave function defined in configuration space. This entails nonlocality. But within wave-functionalism, and under the explanationist approach, the wave function is not a working posit. It is a *presuppositional* one: necessary for the law to function but not something we must commit to ontologically in the same way.

My proposal also fits with the *principle of metaphysical continuity*. According to this principle, the metaphysics we extract from our best theories should be resilient to minor perturbations in formalism, avoiding metaphysical agnosticism about matter. The particle ontology, which remains constant across classical and quantum regimes, offers such a stable metaphysical foundation. In fact, the way I conceive of the nomological ontology is such that it is adaptable across theories precisely in order to preserve the particle ontology—thereby reinforcing its robustness.

I have emphasized *continuity in explanation*, not only in ontology. In classical physics, constructive explanations allow us to derive principle-based theories (like thermodynamics) from deeper dynamical accounts. While empirical data shows that classical mechanics is not the whole story, this does not justify discarding its entire framework. The rational approach, I would think, is to retain the constructive schema and revise only what is necessary. Principle theories are undoubtedly useful, but they do not tell us *why* their principles hold. Constructive theories do: they explain how macroscopic constraints emerge from microscopic dynamics. But such dynamics must describe something concrete. They must be rooted in a microscopic spatiotemporal ontology. Without such an ontology, we would need to postulate additional unexplained principles just to recover spacetime. Moreover, a property-based ontology will not suffice: constructive explanations require a substance ontology. And if we take symmetries seriously—as constraints for perspective-free descriptions—then only a substance ontology can make their explanatory role intelligible.

One might object that my discussion has centered on *non-relativistic quantum mechanics* and therefore might not support sweeping metaphysical conclusions. For instance, some have questioned the viability of a particle ontology in relativistic quantum theories, such as quantum field theory. But quantum field theories are increasingly interpreted as *effective theories*, not as fundamental descriptions. And importantly, there exist particle-based theories that reproduce the empirical predictions of quantum field theories. I should stress that I am not dogmatically attached to the particle ontology. If future evidence or developments provide strong reasons to abandon it, we should follow where the evidence leads. But for now, I see no compelling reason to give it up—especially given how effectively it works.

Suppose, however, that fields ultimately turn out to be fundamental. Would this undermine my entire framework? I do not think so. If constructive explanation remains our guiding goal, then even a field-based fundamental theory would need to explain why, at higher levels, fields behave like particles. This would be a modern realization of Schrödinger's vision of a wave ontology—particles as wave packets—without its conceptual pitfalls. Would this threaten metaphysical continuity? No more than any other empirical course correction. I have never claimed to provide an *infallible methodology* for naturalized metaphysics. It is, like science itself, a process of trial and error. Currently, the most resilient metaphysical picture is particle-based. If that changes, so must we. Agnosticism about substance and a focus solely on properties may appear more cautious, but it also reduces explanatory power. After all, if properties exist, what are they *properties of*?

The *real challenge*, I believe, is not internal to quantum theory but lies in its relationship to relativity. Einstein argued that quantum theory is incomplete because it is nonlocal. I think we should instead conclude that *relativity is incomplete*, because it cannot accommodate quantum nonlocality—a feature confirmed by empirical findings such as those from the Aspect experiments. Our deep-seated preference for locality has been strongly, almost decisively, undermined. There are alternatives to accepting nonlocality, such as superdeterminism and retrocausality, but whether these approaches are worth the theoretical costs is far from clear, and should be further investigated.

This, I believe, is where future research should focus: *How can we constructively explain nonlocality? Do alternatives like superdeterminism or retrocausality sacrifice too much to avoid nonlocality? Can they preserve the virtues of constructive explanation?*

To summarize and conclude: the framework I have presented offers a coherent and promising approach to quantum ontology. It is grounded in the principles of constructive explanation and respect for symmetries. By emphasizing continuity between classical and quantum theories, I have argued that the particle ontology provides a stable metaphysical foundation and a path toward a satisfactory constructive explanation. While the wave function plays a vital role in quantum dynamics, its status is nomological, not ontological. This perspective helps resist the pessimistic meta-induction by maintaining the working posits that have consistently contributed to empirical success. Although relativistic quantum theories raise further challenges—especially regarding nonlocality—I believe the particle ontology still stands as a viable and fruitful foundation for a constructive framework. As quantum field theories are increasingly viewed as effective rather than fundamental, the door remains open to a deeper, but still constructive theory. Ultimately, the most pressing and promising direction for future inquiry is to develop a coherent constructive account of nonlocality—without appealing to radical departures from the explanatory framework that has served physics so well.

References

Adlam, Emily. 2021. *Foundations of Quantum Mechanics*. Cambridge University Press.

Adlam, Emily. 2022. "Two Roads to Retrocausality." *Synthese* 200 (5): 1–36.

Aharonov, Yakir, David Z. Albert, and Lev Vaidman. 1988. "How the Result of a Measurement of a Component of the Spin of a Spin-1/2 Particle Can Turn Out to Be 100." *Physical Review Letters* 60 (14): 1351–4.

Aharonov, Yakir, Peter G. Bergmann, and Joel L. Lebowitz. 1964. "Time Symmetry in the Quantum Process of Measurement." *Physical Review. American Physical Society* 134 (6B): B1410–B1416.

Ainsworth, Peter. 2010. "What Is Ontic Structural Realism?" *Studies in History and Philosophy of Science Part B* 41 (1): 50–7.

Albert, David Z. 1992. *Quantum Mechanics and Experience*. Harvard University Press.

Albert, David Z. 1996. "Elementary Quantum Metaphysics." In: J. Cushing, A. Fine, and S. Goldstein (eds.) *Bohmian Mechanics and Quantum Theory: An Appraisal*. 184: 277–84. Boston Studies in the Philosophy of Science.

Albert, David Z. 2000. *Time and Chance*. Harvard University Press.

Albert, David Z. 2015. *After Physics*. Harvard University Press.

Albert, David Z., and Alyssa Ney. 2013. *The Wave Function: Essays on the Metaphysics of Quantum Mechanics*. Oxford University Press.

Allen, Sophie. 2016. *A Critical Introduction to Properties*. Bloomsbury Publishing.

Allen, Sophie. 2018. "Properties." In: J. Fieser, and B. Dowden (eds.) *Internet Encyclopedia of Philosophy*. https://iep.utm.edu/properties/

Allori, Valia. 2013a. "Primitive Ontology and the Structure of Fundamental Physical Theories." In: D. Albert, and A. Ney (eds.) *The Wave Function: Essays in the Metaphysics of Quantum Mechanics*: 58–75. Oxford University Press.

Allori, Valia. 2013b. "On the Metaphysics of Quantum Mechanics." In: S. Lebihan (ed.) *Precis de la Philosophie de la Physique*: 116–51. Vuibert le Bihan.

Allori, Valia. 2015a. "Maxwell's Paradox: The Metaphysics of Classical Electrodynamics and Its Time-Reversal Invariance." *αnalytica* 1 (1): 1–19.

Allori, Valia. 2015b. "Quantum Mechanics and Paradigm Shifts." *Topoi* 32 (2): 313–23.

Allori, Valia. 2015c. "Primitive Ontology in a Nutshell." *International Journal of Quantum Foundations* 1 (3): 107–22.

Allori, Valia. 2016. "Primitive Ontology and the Classical World." In: R. Kastner, J. Jeknic-Dugic, and G. Jaroszkiewicz (eds.) *Quantum Structural Studies: Classical Emergence from the Quantum Level*: 175–99. World Scientific.

Allori, Valia. 2018a. "Primitive Ontology and Scientific Realism. Or: The Pessimistic Meta-Induction and the Nature of the Wave Function." *Lato Sensu: Revue de la Société de Philosophie des Sciences* 5 (1): 69–76.

Allori, Valia. 2018b. "A New Argument for the Nomological Interpretation of the Wave Function: The Galilean Group and the Classical Limit of Nonrelativistic Quantum Mechanics." *International Studies in the Philosophy of Science* 31 (2): 177–88.

Allori, Valia. 2019a. "Quantum Mechanics, Time and Ontology." *Studies in History and Philosophy of Modern Physics* 66: 145–54.
Allori, Valia. 2019b. "Scientific Realism without the Wave-Function." In: J. Saatsi, and S. French (eds.) *Scientific Realism and the Quantum*: 212–28. Oxford University Press.
Allori, Valia. 2020a. "Some Reflections on the Statistical Postulate: Typicality, Probability and Explanation between Deterministic and Indeterministic Theories." In: V. Allori (ed.) *Statistical Mechanics and Scientific Explanation: Determinism, Indeterminism, and Laws of Nature*: 65–111. World Scientific.
Allori, Valia. 2020b. "Spontaneous Localization Theories with a Particle Ontology." In: V. Allori, A. Bassi, D. Dürr, and N. Zanghì (eds.) *Do Wave Functions Jump? Perspectives on the Work of GianCarlo Ghirardi*: 73–93. Springer.
Allori, Valia. 2021a. "Primitive Beable Is Not Local Ontology: On the Relation between Primitive Ontology and Local Beables." *Crítica* 53 (159): 15–43.
Allori, Valia. 2021b. "Wave-Functionalism." *Synthese* 199: 12271–93.
Allori, Valia. 2022a. "The Paradox of Deterministic Probabilities." *Inquiry: An Interdisciplinary Journal of Philosophy* 67 (10): 3750–69. DOI: 10.1080/0020174X.2022.2065530.
Allori, Valia. 2022b. "Towards a Structuralist Elimination of Properties." In: V. Allori (ed.) *Quantum Mechanics and Fundamentality: Naturalizing Quantum Theory between Scientific Realism and Ontological Indeterminacy*: 141–55. Springer: Synthese Library.
Allori, Valia. 2022c. "Spontaneous Localization Theories." In: O. Freire Jr, G. Bacciagaluppi, O. Darrigol, T. Hartz, C. Joas, A. Kojevnikov, and O. Pessoa Jr (eds.) *Oxford Handbook on the History of Interpretations of Quantum Mechanics*: 1103–30. Oxford University Press.
Allori, Valia. 2022d. "What Is It Like to Be a Relativistic GRW Theory? Or: Quantum Mechanics and Relativity, Still in Conflict after All These Years." *Foundations of Physics* 52: 79.
Allori, Valia. 2022e. "On the Galilei Invariance of the Pilot-Wave Theory." *Foundations of Physics* 52: 111.
Allori, Valia. 2023a. "Who's Afraid of the Measurement Problem?" In: A. Bassi, S. Goldstein, R. Tumulka, and N. Zanghì (eds.) *Physics and the Nature of Reality: Essays in Memory of Detlef Dürr*: 393–409. Springer.
Allori, Valia. 2023b. "What if We Lived in the Best of All Possible (Quantum) Worlds?" In: P. Castro, J. W. M. Bush, and J. R. Croca (eds.) *Advances in Pilot Wave Theory – From Experiments to Foundations*: 227–55. Boston Studies in the Philosophy and History of Science. Springer.
Allori, Valia. 2023c. "Many-Worlds: Why Is It Not the Consensus?" *Quantum Reports* 5 (1): 80–101.
Allori, Valia. 2023d. "Fundamental Objects without Fundamental Properties." In: D. Aerts, J. Arenhart, C. De Ronde, and G. Sergioli (eds.) *Probing the Meaning and Structure of Quantum Mechanics*: 138–67. World Scientific.
Allori, Valia. 2024a. "Hidden Variables and Bell's Theorem: Local or Not?" *Theoria* 39 (2): 143–63. DOI: 10.1387/theoria.24617.
Allori, Valia. 2024b. "Relativistic Pilot-Wave Theories as the Rational Completion of Quantum Mechanics and Relativity." In: A. Oldofredi (ed.) *Guiding Waves in Quantum Mechanics—One Hundred Years of de Broglie-Bohm Pilot-Wave Theory*: 175–92. Oxford University Press.
Allori, Valia. 2025. "Time for Pancakes." In: O. Lombardi and C. Lopez (eds.) *The Arrow of Time: From Closed Systems to the Whole Universe*: 161–83. Cambridge University Press.

Allori, Valia. 2026. " On Einstein's Statistical Interpretation and the de Broglie-Bohm Pilot-wave Theory." Manuscript.
Allori, Valia, and Nino Zanghì. 2004. "What Is Bohmian Mechanics." *International Journal of Theoretical Physics* 43: 1743–55.
Allori, Valia, and Nino Zanghì. 2009. "On the Classical Limit of Quantum Mechanics." *Foundations of Physics* 39 (1): 20–32.
Allori, Valia, Angelo Bassi, Detlef Dürr, and Nino Zanghì. 2020. *Do Wave Functions Jump? Perspectives on the Work of GianCarlo Ghirardi.* Springer.
Allori, Valia, Sheldon Goldstein, Detlef Dürr, and Nino Zanghì. 2002. "Seven Steps toward the Classical World." *Journal of Optics B* 4: 482–8.
Allori, Valia, Sheldon Goldstein, Roderich Tumulka, and Nino Zanghì. 2008. "On the Common Structure of Bohmian Mechanics and the Ghirardi-Rimini-Weber Theory." *The British Journal for the Philosophy of Science* 59 (3): 353–89.
Allori, Valia, Sheldon Goldstein, Roderich Tumulka, and Nino Zanghì. 2011. "Many-Worlds and Schrödinger's First Quantum Theory." *The British Journal for the Philosophy of Science* 62 (1): 1–27.
Allori, Valia, Sheldon Goldstein, Roderich Tumulka, and Nino Zanghì. 2014. "Predictions and Primitive Ontology in Quantum Foundations: A Study of Examples." *The British Journal for the Philosophy of Science* 65 (2): 323–52.
Anandan, Jeeva, and Yakir Aharonov. 1999. "The Meaning of the Density Matrix." *Foundations of Physics Letters* 12 (6): 571–8.
Anderson, James L. 1967. *Principles of Relativity Physics*. Academic Press.
Armstrong, David. 1983. *What Is a Law of Nature?* Cambridge University Press.
Arntzenius, Franck. 2004. "Time-Reversal Operation, Representation of the Lorentz Group and the Direction of Time." *Studies in Histories and Philosophy of Modern Physics* 35 (1): 31–43.
Asistarhov, Serj, and Siddhant Das. 2024. "Comment on 'the Spin Dependence of Detection Times and the Nonmeasurability of Arrival Times.'" DOI: 10.48550/arXiv.2312.01802.
Aspect, Alain, Jean Dalibard, and Gérard Roger. 1982. "Experimental Test of Bell's Inequalities Using Time-Varying Analyzers." *Physical Review Letters* 49 (25): 1804–7.
Baas, Augustine, and Baptiste LeBihan. 2021. "What Does the World Look Like According to Superdeterminism?" *The British Journal for the Philosophy of Science* 74 (3). DOI: 10.1086/714815.
Bacciagaluppi, Guido. 2020."The Role of Decoherence in Quantum Mechanics." In: Edward N. Zalta (ed.) *The Stanford Encyclopedia of Philosophy*. https://plato.stanford.edu/archives/fall2020/entries/qm-decoherence.
Bacciagaluppi, Guido, and Antony Valentini. 2009. *Quantum Theory at the Crossroads: Reconsidering the 1927 Solvay Conference*. Cambridge University Press.
Bacciagaluppi, Guido, and Elise Crull (eds.). 2016. *Grete Hermann—Between Physics and Philosophy*. Spinger.
Badino, Massimiliano. 2020. "Reassessing Typicality Explanation in Statistical Mechanics." In: V. Allori (ed.) *Statistical Mechanics and Scientific Explanation: Determinism, Indeterminism and Laws of Nature*: 147–72. World Scientific.
Baker, Alan. 2003. "Quantitative Parsimony and Explanation." *British Journal for the Philosophy of Science* 54: 245–59.
Baker, Alan. 2007. "Occam's Razor in Science: A Case Study from Biogeography." *Biology and Philosophy* 22 (2): 193–215. DOI: 10.1007/s10539-006-9027-9.

Baker, Alan. 2022. "Simplicity." In: Edward N. Zalta (ed.) *The Stanford Encyclopedia of Philosophy*. https://plato.stanford.edu/archives/sum2022/entries/simplicity.
Baker, David. 2010. "Symmetry and the Metaphysics of Physics." *Philosophy Compass* 5 (12): 1157–66.
Balashov, Yuri, and Michel Janssen. 2003. "Presentism and Relativity." *British Journal for the Philosophy of Science* 54 (2): 327–46.
Barrett, Jeffrey A. 1999. *The Quantum Mechanics of Minds and Worlds*. Oxford University Press.
Barrett, Jeffrey A. 2018. "Everett's Relative-State Formulation of Quantum Mechanics." In: Edward N. Zalta (ed.) *The Stanford Encyclopedia of Philosophy*. https://plato.stanford.edu/archives/win2018/entries/qm-everett.
Barrett, Jeffery A. 2020. *The Conceptual Foundations of Quantum Mechanics*. Oxford University Press.
Barrett, Jeffrey A., and Peter Byrne (eds.). 2012. *The Everett Interpretation of Quantum Mechanics: Collected Works 1955–1980 with Commentary*. Princeton University Press.
Barrett, Thomas. 2015. "On the Structure of Classical Mechanics." *British Journal for the Philosophy of Science* 66 (4): 801–28.
Bassi, Angelo. 2007. "Dynamical Reduction Models: Present Status and Future Developments." *Journal of Physics: Conference Series* 67: 012013.
Bassi, Angelo. 2021."Present and Future Precision Tests of Spontaneous Wave Function Collapse Models." Proc. SPIE 11700, *Optical and Quantum Sensing and Precision Metrology*, 117000S.
Bassi, Angelo, and GianCarlo Ghirardi. 2020. "Collapse Theories." In: Edward N. Zalta (ed.) *The Stanford Encyclopedia of Philosophy*, URL =&#003C; https://plato.stanford.edu/entries/qm-collapse.
Bassi, Angelo, Kinjalk Lochan, Seema Satin, Tejinder P. Singh, and Hendrik Ulbricht. 2013. "Models of Wave-Function Collapse, Underlying Theories, and Experimental Tests." *Review of Modern Physics* 85 (2): 471–527.
Becker, Adam. 2018. *What Is Real? The Unfinished Quest for the Meaning of Quantum Physics*. Basic Books.
Bedingham, Daniel, Detlef Dürr, GianCarlo Ghirardi, Sheldon Goldstein, Roderich Tumulka, and Nino Zanghì. 2014. "Matter Density and Relativistic Models of Wavefunction Collapse." *Journal of Statistical Physics* 154 (1–2): 623–31.
Bell, John S. 1964. "On the Einstein Podolsky Rosen Paradox." *Physics Physique Fizika* 1 (3): 195–200.
Bell, John S. 1966. "On the Problem of Hidden Variables in Quantum Mechanics." *Reviews of Modern Physics* 38 (3): 447–52. -Written before 1964 paper.
Bell, John S. 1975. "The Theory of Local Beables." Presented at the sixth GIFT Seminar Jaca, June 2—7, 1975. Reprinted in: J. S. Bell, 2004. *Speakable and Unspeakable in Quantum Mechanics*: 52. Cambridge.
Bell, John S. 1982. "On the Impossible Pilot Wave." *Foundations of Physics* 12: 98–9.
Bell, John S. 1987. "Are There Quantum Jumps?" In: C. W. Kilmister (ed.) *Schrödinger. Centenary Celebration of a Polymath*: 41–52. Cambridge University Press.
Bell, Mary, and Shan Gao. 2016. *Quantum Nonlocality and Reality. 50 Years of Bell's Theorem*. Cambridge University Press.
Beller, Mara. 2001. *Quantum Dialogue: The Making of a Revolution*. University of Chicago Press.

Belot, Gordon. 2012. "Quantum States for Primitive Ontologists: A Case Study." *European Journal for the Philosophy of Science* 2 (1): 67–83.
Belousek, Daniel. 1996. "Einstein's 1927 Unpublished Hidden-Variable Theory: Its Background, Context and Significance." *Studies in History and Philosophy of Modern Physics* 27 (4): 437–61.
Benatti, Fabio, Giancarlo Ghirardi, and Renata Grassi. 1995. "Describing the Macroscopic World: Closing the Circle within the Dynamical Reduction Program." *Foundations of Physics* 25: 5–38.
Ben-Menahem, Yemima. 2017. "The PBR Theorem: Whose Side Is It On?" *Studies in History and Philosophy of Modern Physics* 57: 80–8.
Ben-Menahem, Yemima. 2018. "The PBR Theorem and Its Implications." *International Journal of Advances in Science Engineering and Technology* 6 (2): 47–51.
Ben-Menahem, Yemima. 2020. "Pitowsky's Epistemic Interpretation of Quantum Mechanics and the PBR Theorem." In: M. Hemmo, and O. Shenker (eds.) *Quantum, Probability, Logic: The Work and Influence of Itamar Pitowsky*: 101–24. Springer.
Berndl, Karin, Detlef Dürr, Sheldon Goldstein, and Nino Zanghì. 1996. "Nonlocality, Lorentz Invariance, and Bohmian Quantum Theory." *Physical Review* A 53 (4): 2062–73.
Bhogal, Harjit. 2020. "Humeanism about Laws of Nature." *Philosophy Compass* 15 (8): e12696.
Bhogal, Harjit, and Zee Perry. 2017. "What the Humean Should Say about Entanglement." *Noûs* 51 (1): 74–94.
Bird, Alexander, and Emma Tobin. 2023. "Natural Kinds." In: Edward N. Zalta, and Uri Nodelman (eds.) *The Stanford Encyclopedia of Philosophy*. https://plato.stanford.edu/archives/spr2023/entries/natural-kinds.
Bliss, Ricki, and James T. M. Miller. 2021. *The Routledge Handbook of Metametaphysics*. Routledge.
Bliss, Ricki, and Kelly Trogdon. 2021. "Metaphysical Grounding." In: Edward N. Zalta (ed.) *The Stanford Encyclopedia of Philosophy*. https://plato.stanford.edu/archives/win2021/entries/grounding.
Bloch, Felix. 1976. "Heisenberg and the Early Days of Quantum Mechanics." *Physics Today* 29 (12): 23–7.
Bohm, David. 1952. "A Suggested Interpretation of the Quantum Theory in Terms of 'Hidden' Variables, I and II." *Physical Review* 85 (2): 166–93.
Bohm, David, and Basil Hiley. 1993. *The Undivided Universe*. Routledge.
Bohr, Niels. 1928. "The Quantum Postulate and the Recent Development of Atomic Theory." *Nature* 121: 589–90.
Bohr, Niels. 1949. "Discussion with Einstein on Epistemological Problems in Atomic Physics." In: P. A. Schilpp (ed.) *Albert Einstein: Philosopher-Scientist*. 1.
Bokulich, Alisa, and Peter Bokulich (eds.). 2011. *Scientific Structuralism*. Boston Studies in the Philosophy of Science: Volume 281. Springer.
Born, Max. 1926. "Zur Quantenmechanik der Stoßvorgänge." *Zeitschrift für Physik* 37 (12): 863–7. English translation, "On the Quantum Mechanics of Collision Processes." In: J. A. Wheeler, and W. H. Zurek (eds.) (1983): *Quantum Theory and Measurement*: 52–61. Princeton University Press.
Born, Max, Werner Heisenberg, and Pascual Jordan. 1926. "Zur Quantenmechanik II." *Zeitschrift für Physik* 35: 557–615. English translation in van der Waerden (1968), paper 15.

Bowman, Gary E. 2005. "On The Classical Limit in Bohm's Theory." *Foundations of Physics* 35: 605–25.
Boyd, Robert. 1980. "Scientific Realism and Naturalistic Epistemology." In: P. D. Asquith and T. Nickles (eds.) *PSA 1980: Proceedings of the Biennial Meeting of the Philosophy of Science Association* 1: 613–62. Philosophy of Science Association.
Brading, Katherine, and Elena Castellani. 2003. *Symmetries in Physics: Philosophical Reflections*. Cambridge University Press.
Brading, Katherine, Elena Castellani, and Nicholas The. 2021. "Symmetry and Symmetry Breaking." In: Edward N. Zalta (ed.) *The Stanford Encyclopedia of Philosophy*. https://plato.stanford.edu/archives/fall2021/entries/symmetry-breaking.
Bricmont, Jean. 2016. *Making Sense of Quantum Mechanics*. Springer.
Bricmont, Jean. 2022. *Making Sense of Statistical Mechanics*. Springer.
Bricmont, Jean, Sheldon Goldstein, and Douglas L. Hemmick. 2020a. "Schrödinger's Paradox and Proofs of Nonlocality Using Only Perfect Correlations." *Journal of Statistical Physics* 180: 74–91.
Bricmont, Jean, Sheldon Goldstein, and Douglas L. Hemmick. 2020b. "EPR-Bell-Schrödinger Proof of Nonlocality Using Position and Momentum." In: V. Allori, A. Bassi, D. Dürr, and N. Zanghì (eds.) *Do Wave Functions Jump? Perspectives on the Work of G.C. Ghirardi*: 5–33. Springer.
Bricmont, Jean, Sheldon Goldstein, and Douglas L. Hemmick. 2022. "From EPR-Schrödinger Paradox to Nonlocality Based on Perfect Correlations." *Foundations of Physics* 52 (3): 1–14.
Brown, Harvey. 2005. *Physical Relativity: Space-Time Structure from a Dynamical Perspective*. Oxford University Press.
Brown, Harvey, and Oliver Pooley. 2006. "Minkowski Space-Time: A Glorious Non-Entity." In: D. Dieks (ed.) *The Ontology of Spacetime*: 67–89. Elsevier.
Brown, Harvey R., and David Wallace. 2005. "Solving the Measurement Problem: De Broglie-Bohm Loses out to Everett." *Foundations of Physics* 35 (4): 517–40.
Brzović, Zdenka. 2018. "Natural Kinds." In: J. Fieser, and B. Dowden (eds.) *The Internet Encyclopedia of Philosophy*. https://iep.utm.edu/nat-kind.
Bub, Jeffrey, and Itamar Pitowsky. 2010. "Two Dogmas about Quantum Mechanics." In: S. Saunders, J. Barrett. A. Kent, and D. Wallace (eds.) *Many Worlds? Everett, Quantum Theory, and Reality*: 433–59. Oxford University Press.
Busch, Jacob. 2003. "What Structures Could Not Be." *International Studies in the Philosophy of Science* 17 (3): 211–25.
Butterfield, Jeremy. 1992. "Bell's Theorem: What It Takes." *The British Journal for the Philosophy of Science* 43 (1): 41–83.
Byrne, Peter. 2010. *The Many Worlds of Hugh Everett III: Multiple Universes, Mutual Assured Destruction, and the Meltdown of a Nuclear Family*. Oxford University Press.
Callender, Craig. 2015. "One World, One Beable." *Synthese* 192 (10): 3153–77.
Callender, Craig. 2020. "Quantum Mechanics: Keeping It Real?" *The British Journal for the Philosophy of Science* 74 (4): 837–51. DOI: 10.1086/715032.
Calosi, Claudio, and Andrea Oldofredi. 2021. "Relational Quantum Mechanics and the PBR Theorem: A Peaceful Coexistence." *Foundations of Physics* 51 (4): 82.
Carcassi, Gabriele, Andrea Oldofredi, and Christine A. Aidala. 2022. "On the Reality of the Quantum State Once Again." http://philsci-archive.pitt.edu/20165.
Carroll, John W. 2020. "Laws of Nature." In: Edward N. Zalta (ed.) *The Stanford Encyclopedia of Philosophy*. https://plato.stanford.edu/archives/win2020/entries/laws-of-nature.

Carroll, Sean. 2019. *Something Deeply Hidden: Quantum Worlds and the Emergence of Spacetime*. Dutton.
Carroll, Sean. 2022. "Reality as a Vector in Hilbert Space." In: V. Allori (ed.) *Quantum Mechanics and Fundamentality: Naturalizing Quantum Theory between Scientific Realism and Ontological Indeterminacy*: 2011–224. Springer.
Cartwright, Nancy. 1983. *How the Laws of Physics Lie*. Clarendon Press.
Castellani, Elena. 1998. "Galilean Particles: An Example of Constitution of Objects." In: E. Castellani (ed.) *Interpreting Bodies. Classical and Quantum Objects in Modern Physics*: 181–94. Princeton University Press.
Chakravartty, Anjan. 1998. "Semirealism." *Studies in History and Philosophy of Modern Science* 29 (3): 391–408.
Chakravartty, Anjan. 2003. "The Structuralist Conception of Objects." *Philosophy of Science* 70 (5): 867–78.
Chakravartty, Anjan. 2004. "Structuralism as a Form of Scientific Realism." *International Studies in Philosophy of Science* 18 (2&3): 151–71.
Chakravartty, Anjan. 2007. *A Metaphysics for Scientific Realism: Knowing the Unobservable*. Cambridge University Press.
Chakravartty, Anjan. 2017. "Scientific Realism." In: Edward N. Zalta (ed.) *The Stanford Encyclopedia of Philosophy*. https://plato.stanford.edu/archives/sum2017/entries/scientific-realism.
Chang, Hasok. 2003. "Preservative Realism and Its Discontents: Revisiting Caloric." *Philosophy of Science* 70 (5): 902–12.
Chen, Eddy K. 2017. "Our Fundamental Physical Space." *The Journal of Philosophy* 114 (7): 333–65.
Chen, Eddy K. 2018. "Quantum Mechanics in a Time-Asymmetric Universe: On the Nature of the Initial Quantum State." *The British Journal for the Philosophy of Science* 72 (4): 1155–83.
Chen, Eddy K. 2019. "Realism about the Wave Function." *Philosophy Compass* 14 (7): e 12611.
Chen, Eddy K. 2021. "Bell's Theorem, Quantum Probabilities, and Superdeterminism." In: E. Knox, and A. Wilson (eds.) *The Routledge Companion to the Philosophy of Physics*: 300–15. Routledge.
Choi, Sungho, and Michael Fara. 2021. "Dispositions." In: Edward N. Zalta (ed.) *The Stanford Encyclopedia of Philosophy*. https://plato.stanford.edu/archives/spr2021/entries/dispositions.
Ciepielewski, Gerardo, S. Elias Okon, and Daniel Sudarsky. 2021. "On Superdeterministic Rejections of Settings Independence." *The British Journal for the Philosophy of Science* 74 (2). DOI: 10.1086/714819.
Clauser, John F., Michael A. Horne, Abner Shimony, and Richard Holt. 1969. "Proposed Experiment to Test Local Hidden-Variable Theories." *Physical Review Letters* 23 (15): 880–4.
Clifton, Rob, and Hans Halvorson. 2002. "No Place for Particles in Relativistic Quantum Field Theories?" *Philosophy of Science* 69 (1): 1–28.
Crane, Harry, and Isaac Wilhelm. 2020. "The Logic of Typicality." In: V. Allori (ed.) *Statistical Mechanics and Scientific Explanation: Determinism, Indeterminism and Laws of Nature*: 173–230. World Scientific.
Cross, Troy. 2005. "What Is a Disposition?" *Synthese* 144 (3): 321–41. DOI: 10.1007/s11229-005-5857-2.
Crowther, Karen, Niels Linnemann, and Christian Wüthrich. 2021. "Spacetime Functionalism in General Relativity and Quantum Gravity." *Synthese* 199 (Suppl 2): 221–7.

Curiel, Erik. 2014. "Classical Mechanics Is Lagrangian; It Is Not Hamiltonian." *British Journal for the Philosophy of Science* 65 (2): 269–321.
Cushing, James T. 1996. *Quantum Mechanics: Historical Contingency and the Copenhagen Hegemony*. University of Chicago Press.
Cushing, James, and Edward McMullin (eds.). 1989. *Philosophical Consequences of Quantum Theory*. University of Notre Dame Press.
Cushing, James T., Arthur Fine, and Sheldon Goldstein (eds.). 1996. *Bohmian Mechanics and Quantum Theory: An Appraisal*. Boston Studies in the Philosophy of Science, Volume 184. Kluwer Academic Publishers.
Cramer, John G. 1986. "The Transactional Interpretation of Quantum Mechanics." *Reviews of Modern Physics* 58 (3): 647–88.
Crull, Elise. 2013. "Exploring Philosophical Implications of Quantum Decoherence." *Philosophy Compass* 8 (9): 275–85.
Das, Siddhant, and Detlef Dürr. 2019. "Arrival Time Distributions of Spin-1/2 Particles." *Scientific Reports* 9 (1): 2242.
Dasgupta, Shamik. 2016. "Symmetry as an Epistemic Notion (Twice Over)." *British Journal for the Philosophy of Science* 67 (3): 837–78.
Daumer, Martin, Detlef Dürr, Sheldon Goldstein, and Nino Zanghì. 1996. "Naive Realism about Operators." *Erkenntnis* 45 (2–3): 379–97.
de Broglie, Louis. 1924. *Recherches sur la théorie des quanta*. Ph.D. Thesis, Université de Paris. *Annals de Physique* (10) 3: 22–128 (1925). English translation, L. de Broglie (2004): "On the Theory of Quanta," *Foundation of Louis de Broglie* ed.
Dennet, Daniel. 1991. "Real Patterns." *Journal of Philosophy* 88 (1): 27–51.
Deutsch, David. 1997. *The Fabric of Reality: The Science of Parallel Universes and Its Implications*. Allen Lane.
DeWitt, Bryce S. 1970. "Quantum Mechanics and Reality." *Physics Today* 23: 30–5.
DeWitt, Bryce S., and Neill Graham. 1973. *The Many-Worlds Interpretation of Quantum Mechanics*. Princeton University Press.
Diósi, Lajos. 1987. "A Universal Master Equation for the Gravitational Violation of Quantum Mechanics." *Physics Letters A* 120 (8): 377–81.
Diósi, Lajos. 1988. "Quantum Stochastic Processes as Models for State Vector Reduction." *Journal of Physics A: Mathematical and General* 21 (13): 2885–98.
Diósi, Lajos. 1989. "Models for Universal Reduction of Macroscopic Quantum Fluctuations." *Physical Review A* 40 (3): 1165–74.
DiSalle, Robert. 2020. "Space and Time: Inertial Frames." In: Edward N. Zalta (ed.) *The Stanford Encyclopedia of Philosophy*. https://plato.stanford.edu/archives/win2020/entries/spacetime-iframes.
Dorr, Cian. 2022. "Natural Properties." In: Edward N. Zalta (ed.) *The Stanford Encyclopedia of Philosophy*. https://plato.stanford.edu/archives/spr2022/entries/natural-properties.
Dowker, Fay, and Isabelle Herbauts. 2005. "The Status of the Wave Function in Dynamical Collapse Models." *Foundations of Physics Letters* 18: 499–518.
Duhem, Pierre. [1906] 1954. *The Aim and Structure of Physical Theory*. English translation by Philip P. Wiener. Princeton University Press.
Durr, Detlef, and Dustin Lazarovici. 2020. *Understanding Quantum Mechanics: The World According to Modern Quantum Foundations*. Springer.
Durr, Detlef, and Stefan Teufel. 2009. *Bohmian Mechanics: The Physics and Mathematics of Quantum Theory*. Berlin: Springer-Verlag
Dürr, Detlef, Sheldon Goldstein, and Nino Zanghì. 1992. "Quantum Equilibrium and the Origin of Absolute Uncertainty." *Journal of Statistical Physics* 67 (5): 843–907.

Dürr, Detlef, Sheldon Goldstein, and Nino Zanghì. 2004. "Quantum Equilibrium and the Role of Operators as Observables in Quantum Theory." *Journal of Statistical Physics* 116: 959–1055.
Dürr, Detlef, Sheldon Goldstein, Karin Münch-Berndl, and Nino Zanghì. 1999. "Hypersurface Bohm-Dirac Models." *Physical Review* A 60: 2729–36.
Dürr, Detlef, Sheldon Goldstein, Roderich Tumulka, and Nino Zanghì. 2004. "Bohmian Mechanics and Quantum Field Theory." *Physical Review Letters* 93 (9): 1–4.
Dürr, Detlef, Sheldon Goldstein, Roderich Tumulka, and Nino Zanghì. 2005a."On the Role of Density Matrices in Bohmian Mechanics." *Foundations of Physics* 35 (3): 449–67.
Dürr, Detlef, Sheldon Goldstein, Roderich Tumulka, and Nino Zanghì. 2005b. "Bell-Type Quantum Field Theories." *Journal of Physics A: Mathematical and General*. R1–R42.
Dürr, Detlef, Sheldon Goldstein, Travis Norsen, Ward Struyve, and Nino Zanghì. 2014. "Can Bohmian Mechanics Be Made Relativistic?" *Proceedings of the Royal Society* A 470 (2162): 20130699.
Dürr, Detlef, and Dustin Lazarovici. 2020. *Understanding Quantum Mechanics: The World According to Modern Quantum Foundations*. Springer.
Dürr, Detlef, and Stefan Teufel. 2009. *Bohmian Mechanics: The Physics and Mathematics of Quantum Theory*. Springer-Verlag.
Egg, Matthias. 2019. "Dissolving the Measurement Problem Is Not an Option for the Realist." *Studies in History and Philosophy of Science Part B: Studies in History and Philosophy of Modern Physics* 66: 62–8.
Einstein, Albert. 1905. "On a Heuristic Viewpoint Concerning the Production and Transformation of Light." *Annalen der Physik* 17: 132–48.
Einstein, Albert. 1919. "What Is the Theory of Relativity?" *The London Times*: November 28.
Einstein, Albert. 1953. "Elementare überlegungen zur Interpretation der Grundlagen der Quanten-Mechanik." In: Sir Edward Appleton et al. (eds.) *Scientific Papers Presented to Max Born on His Retirement from the Tait Chair on Natural Philosophy in the University of Edinburgh*: 33–40. Oliver and Boyd. English translation, "Elementary Considerations on the Interpretation of the Foundations of Quantum Mechanics." https://arxiv.org/abs/1107.3701.
Einstein, Albert, and Leopold Infeld. 1938. *The Evolution of Physics*. Cambridge University Press.
Einstein, Albert, Boris Podolsky, and Nathan Rosen. 1935. "Can Quantum-Mechanical Description of Physical Reality Be Considered Complete?" *Physical Review* 47 (10): 777–80.
Emery, Nina. 2017. "Against Radical Quantum Ontologies." *Philosophy and Phenomenological Research* 95 (3): 564–91.
Esfeld, Michael A. 2004. "Quantum Entanglement and a Metaphysics of Relations." *Studies in History and Philosophy of Modern Physics* 35 (4): 601–17.
Esfeld, Michael. 2009. "The Modal Nature of Structures in Ontic Structural Realism." *International Studies in the Philosophy of Science* 23 (2): 179–94.
Esfeld, Michael A. 2014a. "Quantum Humeanism, Or: Physicalism without Properties." *The Philosophical Quarterly* 64 (256): 453–70.
Esfeld, Michael. 2014b. "How to Account for Quantum Non-locality: Ontic Structural Realism and the Primitive Ontology of Quantum Physics." *Synthese* 194: 2329–44.
Esfeld, Michael, and Vincent Lam. 2008. "Moderate Structural Realism about Space-Time." *Synthese* 160 (1): 27–46.
Esfeld, Michael, and Vincent Lam. 2011. "Ontic Structural Realism as a Metaphysics of Objects." In: Alisa Bokulich and Peter Bokulic (eds.) *Scientific Structuralism*: 143–59. Springer.

Esfeld, Michael A., Dustin Lazarovici, Vincent Lam, and Mario Hubert. "The Physics and Metaphysics of Primitive Stuff." *The British Journal for the Philosophy of Science* 68 (1): 133–61.
Everett, Hugh III. 1957. "'Relative State' Formulation of Quantum Mechanics." *Reviews of Modern Physics* 29 (3): 454–62.
Fano, Vincenzo, Giovanni Macchia, and Gino Tarozzi. 2019. "Is Einstein's Interpretation of Quantum Mechanics ψ-Epistemic?" *Axiomathes* 29 (6): 607–19.
Felline, Laura. 2011. "Scientific Explanation between Principle and Constructive Theories." *Philosophy of Science* 78 (5): 989–1000.
Feynman, Richard, Robert B. Leighton, and Matthew Sands. 1963. *The Feynman Lectures on Physics, Volume III, Quantum Mechanics*. AbeBooks.
Fine, Arthur. 1986. *The Shaky Game: Einstein, Realism and the Quantum Theory*. University of Chicago Press.
Fine, Arthur, and Thomas A. Ryckman. 2020. "The Einstein-Podolsky-Rosen Argument in Quantum Theory." In: Edward N. Zalta (ed.) *The Stanford Encyclopedia of Philosophy*. https://plato.stanford.edu/archives/sum2020/entries/qt-epr.
Flores, Francisco. 1999. "Einstein's Theory of Theories and Types of Theoretical Explanation." *International Studies in the Philosophy of Science* 13 (2): 123–34.
Forrest, Peter. 1988. *Quantum Metaphysics*. Blackwell.
Freedman, Stuart J., and John F. Clauser. 1972. "Experimental Test of Local Hidden-Variable Theories." *Physical Review Letters* 28 (938): 938–41.
Freire, Olival Jr. 2015. *The Quantum Dissidents: Rebuilding the Foundations of Quantum Mechanics (1950–1990)*. Springer.
Freire, Olival Jr., Guido Bacciagaluppi, Guido Bacciagaluppi, Olivier Darrigol, Thiago Hartz, Christian Joas, Alexei Kojevnikov, and Osvaldo Pessoa Jr. (eds.). 2022. *Oxford Handbook on the History of Interpretations of Quantum Mechanics*. Oxford University Press.
French, Stephen. 1989. "Identity and Individuality in Classical and Quantum Physics." *Australasian Journal of Philosophy* 67 (4): 432–46.
French, Stephen. 2013. "Whither Wave-Function Realism?" In: D. Z. Albert, and A. Ney (eds.) *The Wave Function: Essays on the Metaphysics of Quantum Mechanics*: 61–90. Oxford University Press.
French, Stephen. 2014. *The Structure of the World: Metaphysics and Representation*. Oxford University Press.
French, Stephen, and Decio Krause. 2006. *Identity in Physics: A Historical, Philosophical and Formal Analysis*. Oxford University Press.
French, Stephen, and James Ladyman. 2003. "Remodelling Structural Realism: Quantum Physics and the Metaphysics of Structure." *Synthese* 136: 31–56.
French, Stephen, and James Ladyman. 2011. "In Defence of Ontic Structural Realism." In: A. Bokulich, and P. Bokulich (eds.) *Scientific Structuralism*: 25–42. Springer.
French, Stephen, and Michael Redhead. 1988. "Quantum Physics and the Identity of Indiscernibles." *The British Journal for the Philosophy of Science* 39 (2): 233–46.
French, Steven. 1998. "On the Withering Away of Physical Objects." In: E. Castellani (ed.) *Interpreting Bodies: Classical and Quantum Objects in Modern Physics*: 93–11. Princeton University Press.
Friederich, Simon, and Peter W. Evans. 2022. "Retrocausality in Quantum Mechanics." In: Edward N. Zalta, and Uri Nodelman (eds.) *The Stanford Encyclopedia of Philosophy*. https://plato.stanford.edu/archives/win2022/entries/qm-retrocausality.
Friedman, Michael. 1983. *Foundations of Space-time Theories: Relativistic Physics and Philosophy of Science*. Princeton University Press.

Frigg, Roman. 2009. “Typicality and the Approach to Equilibrium in Boltzmannian Statistical Mechanics.” *Philosophy of Science* 76 (5): 997–1008.
Frigg, Roman. 2011. “Why Typicality Does Not Explain the Approach to Equilibrium.” In: M. Suárez (ed.) *Probabilities, Causes and Propensities in Physics*: 7793. Springer.
Frigg, Roman, and Charlotte Werndl. 2012. “Demystifying Typicality.” *Philosophy of Science* 79 (5): 917–29.
Frigg, Roman, and Ioannis Votsis. 2011. “Everything You Always Wanted to Know about Structural Realism but Were Afraid to Ask.” *European Journal for Philosophy of Science* 1 (2): 227–76.
Frisch, Mathias. 2008. “Conceptual Problems in Classical Electrodynamics.” *Philosophy of Science* 75 (1): 93–105.
Frisch, Mathias. 2009. “Philosophical Issues in Electromagnetism.” *Philosophy Compass* 4 (1) (2009): 255–70.
Fuchs, Christopher A., N. David Mermin, and Rüdiger Schack. 2014. “An Introduction to QBism with an Application to the Locality of Quantum Mechanics.” *American Journal of Physics* 82 (8): 749–52.
Gao, Shan (ed.). 2018. *Collapse of the Wave Function: Models, Ontology, Origin, and Implications*. Cambridge University Press.
Ghirardi, GianCarlo. 2012. “Does Quantum Nonlocality Irremediably Conflict with Special Relativity?” *Foundations of Physics* 40: 1379–95.
Ghirardi, GianCarlo, and Angelo Bassi. 2003. “Dynamical Reduction Models.” *Physics Reports* 379 (5–6): 257–426.
Ghirardi, GianCarlo, and Angelo Bassi. 2020. “Collapse Theories.” In: Edward N. Zalta (ed.) *The Stanford Encyclopedia of Philosophy*. https://plato.stanford.edu/archives/sum2020/entries/qm-collapse.
Ghirardi, GianCarlo, and Renata Grassi. 1996. “Bohm’s Theory versus Dynamical Reduction.” In: J. Cushing, A. Fine, and S. Goldstein (eds.) *Bohmian Mechanics and Quantum Theory: An Appraisal*: 353–77. Springer.
Ghirardi, GianCarlo, Alberto Rimini, and Tullio Weber. 1986. “Unified Dynamics for Microscopic and Macroscopic Systems.” *Physical Review D* 34 (2): 470–91.
Giovannelli, Marco, and Don A. Howard. 2025. “Einstein’s Philosophy of Science.” In: Edward N. Zalta (ed.) *The Stanford Encyclopedia of Philosophy*. https://plato.stanford.edu/entries/einstein-philscience.
Gisin, Nicolas. 1984. “Quantum Measurements and Stochastic Processes.” *Physical Review Letters* 52 (19): 1657–60.
Gleason, Andrew M. 1957. “Measures on the Closed Subspaces of a Hilbert Space.” *Journal of Mathematics and Mechanics* 6 (4): 885–93.
Goldstein, Sheldon. 2001. “Boltzmann’s Approach to Statistical Mechanics,” In: J. Bricmont, G. C. Ghirardi, D. Dürr, F. Petruccione, M. C. Galavotti, and N. Zanghí (eds.) *Chance in Physics: Foundations and Perspectives. Lecture Notes in Physics* 574: 39–54. Springer.
Goldstein, Sheldon. 2021. “Bohmian Mechanics.” In: Edward N. Zalta (ed.) *The Stanford Encyclopedia of Philosophy*. https://plato.stanford.edu/archives/fall2021/entries/qm-bohm.
Goldstein Sheldon, and Nino Zanghì. 2013. “Reality and the Role of the Wave Function in Quantum Theory.” In: A. Ney, and D. Albert (eds.) *The Wave Function*: 96–109. Oxford University Press.

Goldstein, Sheldon, and Roderich Tumulka. 2003. "Opposite Arrows of Time Can Reconcile Relativity and Nonlocality." *Classical and Quantum Gravity* 20 (3): 557–64.
Goldstein, Sheldon, and Stefan Teufel. 2001. "Quantum Spacetime without Observers: Ontological Clarity and the Conceptual Foundations of Quantum Gravity." In: C. Callender, and N. Huggett (eds.) *Physics Meets Philosophy at the Planck Scale: Contemporary Theories in Quantum Gravity*: 275–89. Cambridges University Press.
Goldstein, Sheldon, Rodetich Tumulka, and Nino Zanghì. 2024. "On the Spin Dependence of Detection Times and the Nonmeasurability of Arrival Times." *Scientific Reports* 14: 3775. DOI: 10.48550/arXiv.2309.11835.
Goldstein, Sheldon, Travis Norsen, Daniel V. Tausk, and Nino Zanghì. 2011. "Bell's Theorem." *Scholarpedia* 6 (10): 8378.
Goodman, Nelson. 1955. *Fact, Fiction, and Forecast*. Harvard University Press.
Greenberger, Danie, Brigitte Falkenburg, Klaus Hentschel, and Friedel Weinert (eds.). 2009. *Compendium of Quantum Physics: Concepts, Experiments, History and Philosophy*. Spinger.
Hacking, Ian. 1982. "Experimentation and Scientific Realism." *Philosophical Topics* 13 (1): 71–87.
Hall, Ned. 2015. "Humean Reductionism about Laws of Nature." In: B. Loewer, and J. Schaffer (eds.) *A Companion to David Lewis*: 262–77. John Wiley and Sons.
Harker, David. 2013. "How to Split a Theory: Defending Selective Realism and Convergence without Proximity." *British Journal for the Philosophy of Science* 64 (1): 79–106.
Harrigan, Nicholas, and Robert W. Spekkens. 2010. "Einstein, Incompleteness, and the Epistemic View of Quantum States." *Foundations of Physics* 40 (2): 125–57.
Healey, Richard. 2012. "Quantum Theory: A Pragmatist Approach." *British Journal for the Philosophy of Science* 63 (4): 729–71.
Healey, Richard. 2017. *The Quantum Revolution in Philosophy*. Oxford University Press.
Healey, Richard. 2020. "Pragmatist Quantum Realism." In: S. French, and J. Saatsi (eds.) *Scientific Realism and the Quantum*: 123–46. Oxford University Press.
Healey, Richard. 2022. "Quantum-Bayesian and Pragmatist Views of Quantum Theory." In: Edward N. Zalta (ed.) *The Stanford Encyclopedia of Philosophy*. https://plato.stanford.edu/archives/sum2022/entries/quantum-bayesian.
Heisenberg, Werner. 1927. "Ueber die Grundprincipien der 'Quantenmechanik.'" *Forschungen und Fortschritte* 3: 83. Translated in Wheeler and Zurek (1983) as "The Physical Content of Quantum Kinematics and Dynamics" (pp. 62–86).
Heisenberg, Werner. 1949. *The Physical Principles of the Quantum Theory*. Dover.
Hemmick, Douglas L. 1996. *Hidden Variables and Nonlocality in Quantum Mechanics*. Ph. D. thesis, Rutgers. arXiv:quantph/0412011v1.
Hemmick, Douglass L., and Asif M. Shakur. 2012. *Bell's Theorem and Quantum Realism: Reassessment in Light of the Schrödinger Paradox*. Springer.
Hempel, Carl G. 1952. "The Function of General Laws in History." *Journal of Philosophy* 39 (2): 35–48.
Hermann, Grete. 1935. "Die naturphilosophischen Grundlagen der Quantenmechanik." *Abhandlungen der Fries'schen Schule* 6 (2): 75–152.
Hildebrand, Tyler. 2020. "Non-Humean Theories of Natural Necessity." *Philosophy Compass* 15 (5): e12662.
Hildebrand, Tyler. 2023. *Laws of Nature*. Cambridge University Press.

Hilgevoord, Jan, and Jos Uffink. 2016. "The Uncertainty Principle." In: Edward N. Zalta (ed.) *The Stanford Encyclopedia of Philosophy*. https://plato.stanford.edu/archives/win2016/entries/qt-uncertainty.

Hoefer, Carl, Nick Huggett, and James Read. 2023. "Absolute and Relational Space and Motion: Classical Theories." In: Edward N. Zalta & Uri Nodelman (eds.) *The Stanford Encyclopedia of Philosophy*. https://plato.stanford.edu/archives/spr2023/entries/spacetime-theories-classical.

Holland, Peter R. 1993. *The Quantum Theory of Motion*. Cambridge University Press.

Holland, Peter R. 1996. "Is Quantum Mechanics Universal?" In: J. T. Cushing, A. Fine, and S. Goldstein (eds.) *Bohmian Mechanics and Quantum Theory: An Appraisal*: 99–110. Kluwer.

Holland, Peter R. 2005. "What's Wrong with Einstein's 1927 Hidden-Variable Interpretation of Quantum Mechanics?" *Foundations of Physics* 35 (2): 177–96.

Hossenfelder, Sabine, and Thomas Palmer. 2020. "Rethinking Superdeterminism." *Frontiers in Physics* 8: 139.

Howard, Don. 1985. "Einstein on Locality and Separability." *Studies in History and Philosophy of Science* 16 (3): 171–201.

Howard, Don. 1990. "'Nicht sein kann was nicht sein darf', or the Prehistory of EPR, 1909–1935." In: A.I. Miller (ed.) *Sixty-Two Years of Uncertainty*: 61–111. Plenum.

Howard, Don. 2004. "Who Invented the 'Copenhagen Interpretation'? A Study in Mythology." *Philosophy of Science* 71 (5): 669–82.

Howard, Don. 2007. "Revisiting the Einstein-Bohr Dialogue." *Iyyun: The Jerusalem Philosophical Quaterly* 56. *Special Issue Dedicated to the Memory of Mara Beller*: 57–90.

Hubert, Mario. 2021. "Reviving Frequentism." *Synthese* 199: 5255–84.

Hubert, Mario. 2023. "Is the Statistical Interpretation of Quantum Mechanics ψ-Ontic or ψ-Epistemic?" *Foundations of Physics* 53 (16): 1–23.

Hubert, Mario, and Davide Romano. 2018. "The Wave—Function Is a Multi-Field." *European Journal for the Philosophy of Science* 8 (3): 521–37.

Hubert, Mario, and Vera Hartenstein. 2021. "When Fields Are Not Degrees of Freedom." *British Journal for the Philosophy of Science* 72 (1): 245–75.

Huggett, Nick, and Christian Wüthrich. 2025. *Out of Nowhere: The Emergence of Spacetime in Theories of Quantum Gravity*. Oxford University Press.

Ismael, Jenann. 2021. "Quantum Mechanics." In: Edward N. Zalta (ed.) *The Stanford Encyclopedia of Philosophy*. https://plato.stanford.edu/archives/fall2021/entries/qm.

Ismael, Jenann, and Jonathan Schaffer. 2020. "Quantum Holism: Nonseparability as Common Ground." *Synthese* 197: 4131–60. DOI: 10.1007/s11229-016-1201-2.

Jammer, Max. 1974. *The Philosophy of Quantum Mechanics*. Wiley.

Jansson, Lina. 2020. "Can Pragmatism about Quantum Theory Handle Objectivity about Explanations?" In: S. French, and J. Saatsi (eds.) *Scientific Realism and the Quantum*: 147–67. Oxford University Press.

Jarrett, Jon P. 1984. "On the Physical Significance of the Locality Conditions in the Bell Arguments." *Nôus* 18 (4): 569–89.

Joos, Erich, Dieter Zeh, Claus Kiefer, Domenico Giulini, Joachim Kupsch, and Ion-Olimpiu Stamatescu. 2003. *Decoherence and the Appearance of a Classical World in Quantum Theory*. Spinger.

Kastner, Ruth E. 2022. *The Transactional Interpretation of Quantum Mechanics: A Relativistic Treatment (Revised)*. Cambridge University Press.

Kitcher, Philip. 1981. "Explanatory Unification." *Philosophy of Science* 48 (4): 507–31.

Kitcher, Philip. 1993. *The Advancement of Science: Science without Legend, Objectivity without Illusions*. Oxford University Press.
Kitcher, Philip. 2001. "The Galilean Strategy." *The Philosophical Review* 110 (2): 151–97.
Knox, Eleanor. 2013. "Effective Spacetime Geometry." *Studies in History and Philosophy of Modern Physics* 44 (3): 346–56.
Knox, Eleanor. 2014. "Newtonian Spacetime Structure in Light of the Equivalence Principle." *The British Journal for the Philosophy of Science* 65 (4): 863–80.
Knox, Eleanor. 2018. "Physical Relativity from a Functionalist Perspective." *Studies in History and Philosophy of Science Part B: Studies in History and Philosophy of Modern Physics* 67: 118–24.
Kochen, Simon, and Erns P. Specker. 1967. "The Problem of Hidden Variables in Quantum Mechanics." *Journal of Mathematics and Mechanics* 17 (1): 59–87.
Kocsis, Sacha, Boris Braverman, Sylvain Ravets, Martin J. Stevens, Richard P. Mirin, L. Krister Shalm, and Aephraim M. Steinberg. 2011. "Observing the Average Trajectories of Single Photons in a Two-Slit Interferometer." *Science* 332 (6034): 1170–3.
Ladyman, James. 1998. "What Is Structural Realism?" *Studies in History and Philosophy of Science* 29 (3): 409–24.
Ladyman, James. 2008. "Structural Realism and the Relationship between the Special Sciences and Physics." *Philosophy of Science* 75 (5): 744–55.
Ladyman, James. 2020. "Structural Realism." In: Edward N. Zalta (ed.) *The Stanford Encyclopedia of Philosophy*. https://plato.stanford.edu/archives/win2020/entries/structural-realism.
Ladyman, James, and Don Ross. 2007. *Every Thing Must Go: Metaphysics Naturalized*. Oxford University Press.
Lakatos, Imre. 1970. "Falsification and the Methodology of Scientific Research Programmes." In: I. Lakatos, and A. Musgrave (eds.) *Criticism and the Growth of Knowledge*: 91–196. Cambridge University Press.
Lam, Vincent, and Christian Wüthrich. 2018. "Spacetime Is as Spacetime Does." *Studies in History and Philosophy of Modern Physics* 64: 39–51.
Landry, Elaine, and Dean Rickles (eds.). 2012. *Structural Realism: Structure, Object, and Causality*. Springer.
Laudan, Larry. 1981. "A Confutation of Convergent Realism." *Philosophy of Science* 48 (1): 19–49.
Laudisa, Federico, and Carlo Rovelli. 2021. "Relational Quantum Mechanics." In: Edward N. Zalta (ed.) *The Stanford Encyclopedia of Philosophy*. https://plato.stanford.edu/archives/win2021/entries/qm-relational.
Lazarovici, Dustin. 2018. "Against Fields." *European Journal for Philosophy of Science* 8 (2): 145–70.
Lazarovici, Dustin. 2020. *Typicality as a Way of Reasoning in Physics and Metaphysics*. Ph. D. Thesis, University of Lausanne.
Lazarovici, Dustin, and Paula Reichert. 2015. "Typicality, Irreversibility and the Status of Macroscopic Laws." *Erkenntnis* 80 (4): 689–716.
Leifer, Matthew S. 2014. "Is the Quantum State Real? An Extended Review of ψ-Ontology Theorems." *Quanta* 3 (1): 67–155.
Leplin, Jarrett. 1997. *A Novel Defense of Scientific Realism*. Oxford University Press.
Levin, Janet. 2023. "Functionalism." In: Edward N. Zalta and Uri Nodelman (eds.) *The Stanford Encyclopedia of Philosophy*. https://plato.stanford.edu/entries/functionalism/.
Lewis, Peter J. 2001. "Why the Pessimistic Induction Is a Fallacy." *Synthese* 129 (3): 371–80.

Lewis, Peter J. 2004. "Life in Configuration Space." *The British Journal for the Philosophy of Science* 55 (4): 713–29.
Lewis, Peter J. 2005. "Interpreting Spontaneous Collapse Theories." *Studies in History and Philosophy of Modern Physics* 36: 165–80.
Lewis, Peter J. 2006. "GRW: A Case Study in Quantum Ontology." *Philosophy Compass* 1 (2): 224–44.
Lewis, Peter J. 2013. "Dimension and Illusion." In: A. Ney, and D. Z. Albert (eds.) *The Wave Function: Essays on the Metaphysics of Quantum Mechanics*: 110–12. Oxford University Press.
Lewis, Peter J. 2015. "Interpretations of Quantum Mechanics." In: J. Fieser, and B. Dowden (eds.), *The Internet Encyclopedia of Philosophy*. https://iep.utm.edu/int-qm/.
Lewis, Peter J. 2016. *Quantum Ontology: A Guide to the Metaphysics of Quantum Mechanics*. Oxford University Press.
Lewis, Peter J. 2020. "Quantum Mechanics and Its (Dis)Contents." In: S. French, and J. Saatsi (eds.) *Scientific Realism and the Quantum*: 168–82. Oxford University Press.
Lewis, Peter J. 2022. "Collapse Theories." In: E. Knox, and A. Wilson (eds.) *The Routledge Companion to the Philosophy of Physics*: 247–56. Routledge.
Loewer, Barry. 1996. "Humean Supervenience." *Philosophical Topics* 24 (1): 101–27.
Loewer, Barry. 2012. "Two Accounts of Laws and Time." *Philosophical Studies* 160 (1): 115–37.
López, Cristian, and Andrea Oldofredi. 2020. "On the Classification between ψ-Ontic and ψ-Epistemic Ontological Models." *Foundations of Physics* 50 (11): 1315–45.
Lowe, Edward J. 1998. *The Possibility of Metaphysics: Substance, Identity and Time*. Claredon Press.
Lyons, Timothy. 2006. "Scientific Realism and the Stratagema de Divide et Impera." *British Journal for the Philosophy of Science* 57 (3): 537–60.
Lyons, Timothy. 2016. "Structural Realism versus Deployment Realism: A Comparative Evaluation." *Studies in the History and Philosophy of Science* 59: 95–105.
Malament, David. 2004. "On the Time-Reversal Invariance of Classical Electromagnetic Theory." *Studies in Histories and Philosophy of Modern Physics* 35B: 295–315.
Maroney, Owen. 2005. "The Density Matrix in the de Broglie–Bohm Approach." *Foundations of Physics* 35 (3): 493–510.
Marshall, Dan, and Brian Weatherson. 2018. "Intrinsic vs. Extrinsic Properties." In: Edward N. Zalta (ed.) *The Stanford Encyclopedia of Philosophy*. https://plato.stanford.edu/archives/spr2018/entries/intrinsic-extrinsic.
Maudlin, Tim. 1994. *Quantum Non-Locality and Relativity: Metaphysical Intimations of Modern Physics*. Blackwell.
Maudlin, Tim. 1995. "Three Measurements Problems." *Topoi* 14: 7–15.
Maudlin, Tim. 2007a. *The Metaphysics within Physics*. Oxford University Press.
Maudlin, Tim. 2007b. "Completeness, Supervenience, and Ontology." *Journal of Physics A* 40 (12): 3151–71.
Maudlin, Tim. 2007c. "Why Be Humean?" In: Tim Maudlin, *The Metaphysics within Physic*: 50–77. Oxford University Press.
Maudlin, Tim. 2013. "The Nature of the Quantum State." In: A. Ney, and D.Z. Albert (eds.) *The Wave Function: Essays on the Metaphysics of Quantum Mechanics*: 126–53. Oxford University Press.
Maudlin, Tim. 2019a. *Philosophy of Physics: Quantum Theory*. Princeton University Press.
Maudlin, Tim. 2019b. *Bell's Other Assumption(s)*. https://www.youtube.com/watch?v=9VFrDcHnk2c.

Maudlin, Tim. 2020. "The Grammar of Typicality." In: V. Allori (ed.) *Statistical Mechanics and Scientific Explanation: Determinism, Indeterminism and Laws of Nature*: 231–52. World Scientific.
McKenzie, Kerry. 2017. "Ontic Structuralism Realism." *Philosophy Compass* 12 (4): e12399.
Mermin, N. David. 1993. "Hidden Variables and the Two Theorems of John Bell." *Review of Modern Physics* 65 (3): 803–15.
Miller, Elizabeth. 2014. "Quantum Entanglement, Bohmian Mechanics, and Humean Supervenience." *Australasian Journal of Philosophy* 92 (3): 567–83.
Monton, Bradley. 2002. "Wave Function Ontology." *Synthese* 130 (2): 265–77.
Monton, Bradley. 2006. "Quantum Mechanics and 3N—Dimensional Space." *Philosophy of Science* 73 (5): 778–89.
Monton, Bradley. 2013. "Against 3N-Dimensional Space." In: D.Z. Albert, and A. Ney (eds.) *The Wave-Function: Essays in the Metaphysics of Quantum Mechanics*: 154–67. Oxford University Press.
Monton, Bradley, and Chad Mohler. 2021. "Constructive Empiricism." In: Edward N. Zalta (ed.) *The Stanford Encyclopedia of Philosophy*. https://plato.stanford.edu/archives/sum2021/entries/constructive-empiricism.
Morganti, Matteo. 2011. "Is There a Compelling Argument for Ontic Structural Realism?" *Philosophy of Science* 78 (5): 1165–76.
Musgrave, Alan. 1988. "The Ultimate Argument for Scientific Realism." In: R Nola (ed.) *Relativism and Realism in Science*: 229–52. Kluwer.
Myrvold, Wayne. 2015. "What Is a Wave Function?" *Synthese* 192 (10): 3247–74.
Myrvold, Wayne. 2016. "Lessons of Bell's Theorem: Nonlocality, Yes; Action at a Distance, Not Necessarily." In: S. Gao, and M. Bell (eds.) *Quantum Nonlocality and Reality: 50 Years of Bell's Theorem*: 238–60. Cambridge University Press.
Myrvold, Wayne. 2018. "Ontology for Collapse Theories." In S. Gao (ed.) *The Collapse of the Wave Function*: 97–123. Cambridge University Press.
Myrvold, Wayne. 2019. "Ontology for Relativistic Collapse Theories." In: O. Lombardi, S. Fortin, C. Lòpez, and F. Holik (eds.) *Quantum Worlds: Perspectives on the Ontology of Quantum Mechanics*: 9–31. Cambridge University Press.
Myrvold, Wayne. 2021. "Relativistic Constraints on Interpretations of Quantum Mechanics." In: E. Knox, and A. Wilson (eds.) *The Routledge Companion to Philosophy of Physics*: 99–121. Routledge.
Myrvold, Wayne. 2022. "Philosophical Issues in Quantum Theory." In: Edward N. Zalta, and Uri Nodelman (eds.) *The Stanford Encyclopedia of Philosophy*. https://plato.stanford.edu/archives/fall2022/entries/qt-issues.
Myrvold, Wayne, Marco Genovese, and Abner Shimony. 2021. "Bell's Theorem." In: Edward N. Zalta (ed.) *The Stanford Encyclopedia of Philosophy*. https://plato.stanford.edu/archives/fall2021/entries/bell-theorem.
Newman, Mark. 2010. "Beyond Structural Realism: Pluralist Criteria for Theory Evaluation." *Synthese* 174: 413–43.
Newton-Smith, William H. 1981. *The Rationality of Science*. Routledge & Kegan Paul.
Ney, Alyssa. 2012. "The Status of Our Ordinary Three Dimensions in a Quantum Universe." *Noûs* 46 (3): 525–60.
Ney, Alyssa. 2014. *Metaphysics: An Introduction*. Routledge.
Ney, Alyssa. 2015."Fundamental Physical Ontologies and the Constraint of Empirical Coherence: A Defense of Wave Function Realism." *Synthese* 192 (10): 3105–24.

Ney, Alyssa. 2020. "Finding the World in the Wave Function: Some Strategies for Solving the Macro—Object Problem." *Synthese* 197 (10): 4227–4249. DOI: 10.1007/s11229-017-1349-4.

Ney, Alyssa. 2020. "Separability, Locality, and Higher Dimensions in Quantum Mechanics." In: S. Dasgupta, and B. Weslake (eds.) *Current Controversies in Philosophy of Science*: 75–90. London: Routledge.

Ney, Alyssa. 2021. *The World in the Wave Function: A Metaphysics for Quantum Physics.* Oxford University Press.

Niiniluoto, Ikka. 1999. *Critical Scientific Realism*. Oxford University Press.

Niiniluoto, Ilkka. 2019. "Scientific Progress." In: Edward N. Zalta (ed.) *The Stanford Encyclopedia of Philosophy*. https://plato.stanford.edu/archives/win2019/entries/scientific-progress.

Nikolić, Hrvoie. 2005. "Relativistic Quantum Mechanics and the Bohmian Interpretation." *Foundations of Physics Letters* 18: 549–61.

Nikolić, Hrvoje. 2012. "EPR before EPR: A 1930 Einstein-Bohr Thought Experiment Revisited." *European Journal of Physics* 33 (5): 1089–97. DOI: 10.1088/0143-0807/33.

North, Jill. 2013. "The Structure of the Quantum World." In: A. Ney, and D. Z. Albert (eds.) *The Wave Function: Essays on the Metaphysics of Quantum Mechanics*: 184-202. Oxford University Press.

Norsen, Travis. 2006a. "Bell Locality and the Nonlocal Character of Nature." *Foundations of Physics Letters* 19 (7): 633–55.

Norsen, Travis. 2006b. "EPR and Bell Locality." *AIP Conference Proceedings* 844: 281–93.

Norsen, Travis. 2009. "Local Causality and Completeness: Bell vs. Jarrett." *Foundations of Physics* 39: 273–94.

Norsen, Travis. 2010a. "The Theory of (Exclusively) Local Beables." *Foundations of Physics* 40: 1858–84.

Norsen, Travis. 2010b. "Parameter Independence and Outcome Independence in Dynamical Collapse Theories." Manuscript.

Norsen, Travis. 2011. "Bell's Concept of Local Causality." *American Journal of Physics* 79 (12): 1261–75.

Norsen, Travis. 2014. "The Pilot-Wave Perspective on Spin." *American Journal of Physics* 82 (4): 337–48.

Norsen, Travis. 2015. "Are There Really Two Different Bell's Theorems?" *International Journal of Quantum Foundations* 1: 65–84.

Norsen, Travis. 2016. "Quantum Solipsism and Nonlocality." In: S. Gao, and M. Bell (eds.) *Quantum Nonlocality and Reality: 50 Years of Bell's Theorem*: 204–37. Cambridge University Press.

Norsen, Travis. 2017. *Foundations of Quantum Mechanics: An Exploration of the Physical Meaning of Quantum Theory*. Springer.

North, Jill. 2021a. "Formulations of Classical Mechanics." In: E. Knox and A. Wilson (eds.) *The Routledge Companion to Philosophy of Physics*: 21–32. Routledge.

North, Jill. 2021b. *Physics, Structure, and Reality*. Oxford University Press.

O'Connor, Timothy. 2021. "Emergent Properties." In: Edward N. Zalta (ed.) *The Stanford Encyclopedia of Philosophy*. https://plato.stanford.edu/archives/win2021/entries/properties-emergent.

Orilia, Francesco, and Michele Paolini Paoletti. 2022. "Properties." In: Edward N. Zalta (ed.) *The Stanford Encyclopedia of Philosophy*. https://plato.stanford.edu/archives/spr2022/entries/properties.

Pearle, Philip. 1989. “Combining Stochastic Dynamical State-Vector Reduction with Spontaneous Localization.” *Physical Review A* 39 (5): 2277–89.
Pearle, Philip. 2009. “How Stands Collapse II.” In: W. C. Myrvold, and J. Christian (eds.) *Quantum Reality, Relativistic Causality, and Closing the Epistemic Circle*: 257–92. Springer.
Penrose, Roger. 1996. “On Gravity’s Role in Quantum State Reduction.” *General Relativity and Gravitation* 28 (5): 581–600.
Penrose, Roger. 2014. “On the Gravitization of Quantum Mechanics 1: Quantum State Reduction.” *Foundations of Physics* 44 (5): 557–75.
Peters, Dean. 2014. “What Elements of Successful Scientific Theories Are the Correct Targets for ‘Selective’ Scientific Realism?” *Philosophy of Science* 81 (3): 377–97.
Philippidis, Chris, Christopher Dewdney, and Basil J. Hiley. 1979. “Quantum Interference and the Quantum Potential II.” *Nuovo Cimento* B 52 (1): 15–28.
Poincaré, Henri. [1905] 1952. *Science and Hypothesis*. Dover.
Popper, Karl. 1935. *Logik der Forschung. Zur Erkenntnistheorie der Modernen Naturwissenschaft*. Springer. Translated as *The Logic of Scientific Discovery* (1959).
Psillos, Stathis. 1995. “Is Structural Realism the Best of Both Worlds?” *Dialectica* 49 (1): 15–46.
Psillos, Stathis. 1999. *Scientific Realism: How Science Tracks Truth*. Routledge.
Psillos, Stathis. 2001. “Is Structural Realism Possible?” *Philosophy of Science* 68 (Supplementary Volume): S13–S24.
Psillos, Stathis. 2006. “The Structure, the Whole Structure and Nothing but the Structure?” *Philosophy of Science* 73 (3): 560–70.
Psillos, Stathis. 2022. “Realism and Theory Change in Science.” In: Edward N. Zalta and Uri Nodelman (eds.) *The Stanford Encyclopedia of Philosophy*. https://plato.stanford.edu/archives/fall2022/entries/realism-theory-change.
Pusey, Matthew, Jonathan Barrett, and Terry Rudolph. 2012. “On the Reality of the Quantum State.” *Nature Physics* 8: 475–78.
Putnam, Hilary. 1975. *Mathematics, Matter and Method*. Cambridge University Press.
Putnam, Hilary. 1978. *Meaning and the Moral Sciences*. Routledge.
Przibram, Karl. 1967. *Letters on Wave Mechanics: Schrödinger, Planck, Einstein, Lorentz*. New York: Philosophical Library.
Quine, Willard V. O. 1951. “Two Dogmas of Empiricism.” *The Philosophical Review* 60 (1): 20–43.
Quine, Willard V. O. 1975. “On Empirically Equivalent Systems of the World” *Erkenntnis* 9 (3): 313–28.
Redhead, Michael. 1987. *Incompleteness, Nonlocality, and Realism: A Prolegomenon to the Philosophy of Quantum Mechanics*. Oxford University Press.
Rickles, Dean. 2022. “Supervenience and Determination.” In: J. Fieser, and B. Dowden (eds.) *The Internet Encyclopedia of Philosophy*. https://iep.utm.edu/superven.
Romano, Davide, and Antonio Vassallo. 2021. “The Metaphysics of Decoherence.” *Erkenntnis*: 1–23.
Rosaler, Joshua. 2015. “Is de Broglie–Bohm Theory Specially Equipped to Recover Classical Behavior?” *Philosophy of Science* 82 (5): 1175–87.
Rovelli, Carlo. 1996. “Relational Quantum Mechanics.” *International Journal of Theoretical Physics* 35 (8): 1637–78.
Rowbottom, Darrell. 2019. *The Instrument of Science: Scientific Anti-Realism Revitalised*. Routledge.

Saatsi, Juha. 2008. "Eclectic Realism: The Proof of the Pudding." *Studies in History and Philosophy of Science* 39 (2): 273–6.

Sankey, Howard. 2001. "Scientific Realism: An Elaboration and a Defense." *Theoria* 98: 35–54.

Saunders, Simon. 1993. "Decoherence, Relative States, and Evolutionary Adaptation." *Foundations of Physics* 23 (12): 1553–85.

Saunders, Simon. 1995. "Time, Quantum Mechanics, and Decoherence." *Synthese* 102 (2): 235–66.

Saunders, Simon. 2003. "Structural Realism Again." *Synthese* 136: 127–33.

Saunders, Simon, Jeffrey Barrett, Adrian Kent, and David Wallace (eds.). 2010. *Many Worlds? Everett, Quantum Theory, & Reality*. Oxford University Press.

Schaffer, Jonathan. 2010. "Monism: The Priority of the Whole." *Philosophical Review* 119 (1): 31–76.

Schrödinger, Erwin. 1926. "Quantisierung as Eigenwertproblem." (Quantization as an Eigenvalue Problem) *Annalen der Physik* 4 (79): 361–76; 489–527; 80, S: 437–90; 81, S: 109–40. Reprinted in E. Schrödinger (1927): *Abhandlungen zur Wellenmechanik*, Johann Ambrosius Barth. English translation, E. Schrödinger (1928): *Collected Papers on Wave Mechanics*, Glasgow.

Schrödinger, Erwin. 1928. *Collected Papers on Wave Mechanics*. AMS Chelsea Publishing.

Schrödinger, Erwin. 1935a. "Die gegenwärtige Situation in der Quantenmechanik." *Naturwissenschaften* 23: 807–12; 823–28; 844–9 (1935). English translation: "The Present Situation in Quantum Mechanics," translated by J. D. Trimmer, *Proceedings of the American Philosophical Society* 124: 323–338 (1980).

Schrödinger, Erwin. 1935b. "Discussion of Probability Relations between Separated Systems." *Mathematical Proceedings of the Cambridge Philosophical Society* 31: 555–63.

Schrödinger, Erwin. 1936. "Probability Relations between Separated Systems." *Mathematical Proceedings of the Cambridge Philosophical Society* 32: 446–52.

Sellars, Wilfrid. 1962. "Philosophy and the Scientific Image of Man." In: R. Colodny (ed.) *Frontiers of Science and Philosophy*: 35–78. Pittsburgh University Press.

Shimony, Abner. 1984. "Controllable and Uncontrollable Nonlocality." In: S. Kamefuchi *et al*. (eds.) *Foundations of Quantum Mechanics in the Light of New Technology*: 225–30. Physical Society of Japan.

Shimony, Abner. 1993. *The Search for a Naturalistic World View, vol. 1. Scientific Method and Epistemology, vol. 2, Natural Science and Metaphysics*. Cambridge University Press.

Shimony, Abner, Michael A. Horne, and John F. Clauser. 1976. "Comment on 'The Theory of Local Beables.'" *Epistemological Letters* 13 (1): 1–8.

Sider, Theordore. 2011. *Writing the Book of the World*. Oxford University Press.

Sider, Theodore. 2020. *The Tools of Metaphysics and the Metaphysics of Science*. Oxford University Press.

Silberstein, Michael, W. Mark Stuckey, and Timothy McDevitt. 2018. *Unifying Block Universe Physics and Time as Experienced*. Oxford University Press.

Sklar, Lawrence. 1993. *Physics and Chance: Philosophical Issues in the Foundations of Statistical Mechanics*. Cambridge University Press.

Sober, Elliott. 2015. *Ockham's Razors: A User's Manual*. Cambridge University Press.

Stanford, P. Kyle. 2003. "No Refuge for Realism: Selective Confirmation and the History of Science." *Philosophy of Science* 70 (5): 913–25.

Stanford, P. Kyle. 2006. *Exceeding Our Grasp: Science, History, and the Problem of Unconceived Alternatives*. Oxford University Press.

Stanford, Kyle. 2021. "Underdetermination of Scientific Theory." In: Edward N. Zalta (ed.) *The Stanford Encyclopedia of Philosophy*. https://plato.stanford.edu/archives/win2021/entries/scientific-underdetermination.
Struyve, Ward. 2022. "Time Reversal Invariance and Ontology." *British Journal for the Philosophy of Science*. doi/10.1086/721089.
Suárez, Mauricio. 2015. "Bohmian Dispositions." *Synthese* 192 (10). 3203–28.
Sutherland, Roderick. 2008. "Causally Symmetric Bohm Model." *Studies in History and Philosophy of Modern Physics* 39 (4): 782–805.
Swartz, Norman. 2018. "Laws of Nature." In: J. Fieser, and B. Dowden (eds.) *The Internet Encyclopedia of Philosophy*. https://iep.utm.edu/lawofnat.
Tahko, Tuomas E. 2015. *An Introduction to Metametaphysics*. Cambridge University Press.
Tahko, Tuomas E. 2018. "Fundamentality." In: Edward N. Zalta (ed.) *The Stanford Encyclopedia of Philosophy*. https://plato.stanford.edu/archives/fall2018/entries/fundamentality.
Tahko, Tuomas E., and E. Jonathan Lowe. 2020. "Ontological Dependence." In: Edward N. Zalta (ed.) *The Stanford Encyclopedia of Philosophy*. https://plato.stanford.edu/archives/fall2020/entries/dependence-ontological.
Tegmark, Max. 2007. "The Mathematical Universe." *Foundations of Physics* 38 (2): 101–50. DOI: 10.1007/s10701-007-9186-9.
Teller, Paul. 1989. "Relativity, Relational Holism, and the Bell Inequalities." In: J. T. Cushing, and E. McMullin (eds.) *Philosophical Consequences of Quantum Theory: Reflections on Bell's Theorem*: 208–21. University of Notre Dame Press.
t'Hooft, Gerard. 2016. *The Cellular Automaton Interpretation of Quantum Mechanics*. Spinger.
Tumulka, Roderich. 2006a. "A Relativistic Version of the Ghirardi-Rimini-Weber Model." *Journal of Statistical Physics* 125 (4): 821–40. DOI: 10.1007/s10955.
Tumulka, Roderich. 2006b. "Collapse and Relativity." *AIP Conference Proceedings* 844: 340352.
Tumulka, Roderich. 2020. "A Relativistic GRW Flash Process with Interaction." In: V. Allori, A. Bassi, D. Dürr, and N. Zanghí (eds.) *Do Wavefunctions Jump? Perspectives of the Work of GianCarlo Ghirardi*: 321–48. Springer.
Tumulka, Roderich. 2021. "Bohmian Mechanics." In: E. Knox, and A. Wilson (eds.) *The Routledge Companion to the Philosophy of Physics*: 257–71. Routledge.
Tumulka, Roderich. 2022. *Foundations of Quantum Mechanics*. Springer.
Vaidman, Lev. 2021. "Many-Worlds Interpretation of Quantum Mechanics." In: Edward N. Zalta (ed.) *The Stanford Encyclopedia of Philosophy*. https://plato.stanford.edu/archives/fall2021/entries/qm-manyworlds.
Valentini, Antoni. 2007. "Hidden Variables and the Large-Scale Structure of Space-Time." In: W. L. Craig, and Q. Smith (eds.) *Einstein, Relativity and Absolute Simultaneity*: 125–55. Routledge.
Valentini, Antony. 1992. *On the Pilot-Wave Theory of Classical, Quantum and Subquantum Physics*. Ph.D. Thesis, ISAS, Trieste.
Valentini, Antony. 1997. "On Galilean and Lorentz Invariance in Pilot-Wave Dynamics." *Physics Letters A* 228 (4–5): 215–22.
van Camp, Wesley. 2011. "Principle Theories, Constructive Theories, and Explanation in Modern Physics." *Studies in History and Philosophy of Science Part B: Studies in History and Philosophy of Modern Physics* 42 (1): 23–31.
van der Waerden, B. L. 1968. *Sources of Quantum Mechanics*. Dover.

van Fraassen, Baas. 2006. "Structure: Its Shadow and Substance." *British Journal for the Philosophy of Science* 57 (2): 275–307.
van Fraassen, Bas. 1980. *The Scientific Image*. Oxford University Press.
van Inwagen, Peter, and Meghan Sullivan. 2021. "Metaphysics." In: Edward N. Zalta (ed.) *The Stanford Encyclopedia of Philosophy*. https://plato.stanford.edu/archives/win2021/entries/metaphysics.
Vickers, Peter. 2012. "Historical Magic in Old Quantum Theory?" *European Journal for Philosophy of Science* 2 (1): 1–19.
Vickers, Peter. 2013. "A Confrontation of Convergent Realism." *Philosophy of Science* 80 (2): 189–211.
Vickers, Peter. 2017. "Understanding the Selective Realist Defence against the PMI." *Synthese* 194: 3221–32.
Volchan, Sergio B. 2007. "Probability as Typicality." *Studies in History and Philosophy of Modern Physics* 38 (4): 801–14.
von Neumann, John. 1932. *Mathematische Grundlagen der Quantenmechanik*. Springer Verlag. Translated by Robert T. Beyer (1955): *Mathematical Foundations of Quantum Mechanics*. Princeton University Press.
Wallace, David M. 2003. "Everett and Structure." *Studies in History and Philosophy of Modern Physics* 34 (1): 87–105.
Wallace, David M. 2012. *The Emergent Multiverse: Quantum Theory According to the Everett Interpretation*. Oxford University Press.
Wallace, David M. 2020. "Against Wave Function Realism." In: S. Dasgupta, and B. Weslake (eds.) *Current Controversies in Philosophy of Science*: 63–74. Routledge.
Wallace, David M. and Christopher Timpson. 2010. "Quantum Mechanics on Spacetime I: Spacetime State Realism." *British Journal for the Philosophy of Science* 61 (4): 697–727.
Wasserman, Ryan, David Manley, and David Chalmers. 2009. *Metametaphysics: New Essays on the Foundations of Ontology*. Oxford University Press.
Wharton, Ken, and Huw Price. 2023. "Quantum Mechanics: How the Future Might Influence the Past." *The Conversation*. https://theconversation.com/quantum-mechanics-how-the-future-might-influence-the-past-199426.
Wheeler, John A., and Richard P. Feynman. 1945. "Interaction with the Absorber as the Mechanism of Radiation." *Reviews of Modern Physics* 17 (2–3): 157–81.
Wheeler, John A., and Wojciech Zurek (eds.). 1983. *Quantum Theory and Measurement*. Princeton University Press.
Wigner, Eugene P. 1961. "Remarks on the Mind-Body Question." In: I. J. Good (ed.) *The Scientist Speculates*: 284–302. Heinemann.
Wilce, Alexander. 2021. "Quantum Logic and Probability Theory." In: Edward N. Zalta (ed.) *The Stanford Encyclopedia of Philosophy*. https://plato.stanford.edu/archives/fall2021/entries/qt-quantlog.
Wilhelm, Isaac. 2022. "Typical: A Theory of Typicality and Typicality Explanation." *British Journal for the Philosophy of Science* 73 (2): 561–81. DOI: 10.1093/bjps/axz016.
Wilson, Alastair. 2020. *The Nature of Contingency: Quantum Physics as Modal Realism*. Oxford University Press.
Wilson, Jessica. 2006. "ON Characterizing the Physical." *Philosophical Studies* 131 (1): 61–99.
Wilson, Mark. 2013. "What Is 'Classical Mechanics' Anyway?" In: R. Batterman (ed.) *The Oxford Handbook of Philosophy of Physics*: 43–106. Oxford University Press.

Worrall, John. 1989. "Structural Realism: The Best of Both Worlds?" *Dialectica* 43 (1–2): 99–124.
Wüthrich, Christian, and Nick Huggett. 2025. *Out of Nowhere: The Emergence of Spacetime in Theories of Quantum Gravity*. Oxford University Press.
Zeh, H. Dieter. 1970. "On the Interpretation of Measurement in Quantum Theory." In: *Foundations of Physics* 1: 69–76. Reprinted in: J. A. Wheeler and W. H. Zurek (eds.) (1983): *Quantum Theory and Measurement*: 342–52. Princeton University Press.

Index

Page numbers in *italics* refer to figures and tables

www.ingramcontent.com/pod-product-compliance
Ingram Content Group UK Ltd.
Pitfield, Milton Keynes, MK11 3LW, UK
UKHW020214290926
15470UKWH00001B/183

9 781498 581684